PHYSICS

THE STUDY OF
MATTER & ENERGY FROM
A CHRISTIAN WORLDVIEW

Author: Dr. Dennis Englin

Master Books Creative Team:

Editor: Craig Froman

Design: Jennifer Bauer

Cover Design: Diana Bogardus

Copy Editors:
Judy Lewis
Willow Meek

Curriculum Review:
Kristen Pratt
Laura Welch
Diana Bogardus

First printing: April 2022

Copyright © 2022 by Dr. Dennis Englin and Master Books®. All rights reserved. No part of this book may be used or reproduced in any manner whatsoever without written permission of the publisher, except in the case of brief quotations in articles and reviews. For information write:

Master Books®, P.O. Box 726, Green Forest, AR 72638
Master Books® is a division of the New Leaf Publishing Group, Inc.

ISBN: 978-1-68344-222-6
ISBN: 978-1-61458-794-1 (digital)
Library of Congress Number: 2022933045

Unless otherwise noted, Scripture quotations are from the King James Version of the Bible.

Printed in the United States of America

Please visit our website for other great titles:
www.masterbooks.com

Author Bio:

Dr. Dennis Englin enjoys teaching in the areas of animal biology, vertebrate biology, wildlife biology, organismic biology, biology, and astronomy. Memberships include the Creation Research Society, Southern California Academy of Sciences, Yellowstone Association, and Au Sable Institute of Environmental Studies. Dr. Englin's most recent publications are his *Biology* and *Chemistry* courses, both from Master Books. His research interests are in the area of animal studies. He is a retired Professor of Biology at The Master's University in Santa Clarita, California.

B.A., Westmont College
M.S., California State University, Northridge
Ed.D., University of Southern California

TABLE OF CONTENTS

Chapter 1	Physics, Wisdom, and Science	4
Chapter 2	Speed, Velocity, and Acceleration	14
Chapter 3	Force & Newton's 3 Laws of Motion	22
Chapter 4	Vectors	30
Chapter 5	Gravity	36
Chapter 6	Kinetic Energy, Potential Energy, Momentum, and Power	46
Chapter 7	Work, Machines, and Torque	58
Chapter 8	Rotational Motion	66
Chapter 9	Projectile Motion	74
Chapter 10	Kepler's Laws of Planetary Motion	82
Chapter 11	Heat Energy	90
Chapter 12	Laws of Thermodynamics	98
Chapter 13	Waves	106
Chapter 14	Sound Waves	114
Chapter 15	Musical Sound Waves	122
Chapter 16	Electromagnetism	128
Chapter 17	Light Waves — Electromagnetic Radiation	136
Chapter 18	Lenses — Refraction	144
Chapter 19	Dispersion of Light and Reflection	150
Chapter 20	Electric Circuits 1	158
Chapter 21	Electric Circuits 2	166
Chapter 22	Atoms and Other Tiny Things	174
Chapter 23	Radioactivity	182
Chapter 24	Applications of Radioactivity	190
Chapter 25	Time	198
Chapter 26	The Solar System	206
Chapter 27	The Universe	216
Chapter 28	Cosmology	226
Glossary		234

CHAPTER ONE

PHYSICS, WISDOM, AND SCIENCE

OBJECTIVES

At the conclusion of this lesson the student should have an understanding of

- How God's wisdom is shown in the creation of the universe including the natural laws which form the basis for the study of physics

- The ideas that Aristotle developed through observations and philosophy that were held to for almost 2,000 years

- The contributions of Galileo Galilei with the Scientific Method and measurements that laid a foundation for testing ideas in the development of physics

- The use of measurements in physics — the application of the metric system

- Forming and testing a hypothesis from data and observations from an application of Archimedes' Principle

Curiosity and the Natural World

Have you ever wondered how a fully loaded aircraft carrier can float, but a rock cannot? Have you ever wondered how a plane with hundreds of people can fly? How does food become energy that moves your muscles? When you were first born you probably did not wonder much. But you have made up for it since. Sometimes while growing up we are told not to ask so many questions, though that does not stop us from wondering. To be alive and appreciate God's bountiful creation, we need to keep asking those questions. We are made to be curious and to try to answer our questions. Hopefully, this study will answer some of those questions you have always wondered about and cause you to ask more. Why do you feel colder when wet but warmer after you dry off?

God and Physics

Physics is a study of the natural laws, such as the Law of Gravity, which we live by every waking moment of our lives. In our dreams we may soar above treetops and fly with the birds — but not when we are awake. Don't try it. People knew of the natural laws long before they understood them. Have you ever wondered where they came from or what life would be like without them? Most feel that they are just there as part of the world in which we live. They operate the same way all the time. The law of gravity functions the same way on earth and Mars.

Our universe is not just stuff. We also have gravity, light, electric forces, magnetic forces, nuclear forces, and on and on. They are all measurable and follow exact patterns all the time. The only exceptions are miracles. This is the only reason we can study physics. If the natural laws changed, physics would have to be reinvented every time.

"In the beginning God created the heavens and the earth." Genesis 1:1 states that God spoke the universe into being from nothing. This included the natural laws because the universe had to work from the very beginning. Gravity had to determine the orbits of the planets in the solar system and the stars in galaxies and the galaxies in families of galaxies. The sun, moon, and stars were created on the fourth day of creation, so God created light four days before the sun. The light consisted of electromagnetic radiation, so laws controlling electrical forces and magnetic forces had to be there as well from the beginning.

By His wisdom, God created the universe. Proverbs 8:22–23 states, "The Lord possessed me [wisdom] at the beginning of His way, before His works of old. I have been established from everlasting, from the beginning, before there ever was an earth." God (Father, Son, and Holy Spirit) created by His wisdom a fully functioning universe at the beginning. There was light, gravity, and the other functions that are studied in physics at the moment God gave birth to the universe. From the time of Adam and Eve, we have been trying to understand this fascinating universe in which we live. Can you imagine what the new heavens and earth will be like — even greater and without sin?

Biblical wisdom is often referred to as moral and spiritual wisdom. But in Proverbs 8:22–31, God's wisdom is directed at the creation of the physical universe — putting together the physical aspects of the universe. Physics shows us the results of the physical aspects of God's biblical wisdom. It is true that some of man's ideas do not reflect biblical wisdom, but an objective study of the universe itself can show God's wisdom that He put into His creative work.

A natural law is a description of observations made many times that do not vary. For example, gravity is always the same everywhere. Your weight is the pull of gravity from the earth on your mass. Your weight changes as your mass changes or you move closer or farther from the center of the earth. As you move closer to the center of the earth, earth exerts a stronger force of gravity on you and as you move farther from the center of the earth, you weigh less because the force of gravity from the earth on your mass is less. When Jesus walked on water, He altered the force of gravity so that it would not pull Him below the water. As God, the **laws of nature** only affected Jesus as He submitted to them. When He hung on the Cross, He could have altered gravity so that his weight would not have pulled Him down onto the nails tearing into the flesh of His hands and feet. But He chose not to so that He could suffer, paying the ultimate price for our sin.

Natural laws cannot be altered or disobeyed (except by God and whom He allows); they are applicable throughout the universe and are mathematically precise every time. They point to the wisdom and character of God. Therefore, Albert Einstein stated that the existence of God cannot be denied. It has been said that there are no true atheists — just bitter people.

Even as Christians, we often think and act as if the natural laws came about and operate on their own and the only evidence for God are miracles. This is what the Jewish rulers were saying when they told Jesus to give them a sign (perform a miracle). Even when He raised the dead, they did not accept His divine control over creation.

The role of God in creating and maintaining the natural laws is critical to your understanding of physics. You are not studying things made up by a lot of people, but the very handiwork of God Himself.

> **All things were created through him [Jesus] and for him. And he is before all things, and in him all things consist (Colossians 1:16b–17).**

Early Physics — Aristotle

The Greek philosopher Aristotle (384–322 B.C.) studied the motion of objects using observations and logic. He concluded that motion was caused by force — the push and pull on an object. He wondered why a thrown ball kept moving when it left someone's hand. When it left a person's hand there was no longer a force on the ball. He concluded that when the ball moved through the air, the ball pushed the air causing it to go behind the ball and push it. Today we know that momentum and inertia also play a role. Measurement and experiments were not part of his study; this came later. He also concluded that a heavier object will fall faster than a lighter object — it just seemed to make sense. But later measurements and experiments demonstrated that they will fall at the same rate unless they hit air resistance. Therefore, a parachute falls more slowly than a rock — good thing. Aristotle's teachings were accepted as common sense for almost 2,000 years.

School of Aristotle

It is interesting to note that Aristotle tutored Alexander the Great, who conquered most of the Mediterranean civilization and part of Europe. Daniel 7:6 tells that Daniel had a vision while in exile in Babylon of a leopard that had four heads. The leopard was symbolic of Alexander the Great, and the four heads were four generals that would divide his kingdom after his death. History is "His Story" where all the puzzle pieces are part of a larger whole of God moving through time. While studying physics, you are not just studying a bunch of isolated ideas and equations that only a few people care about. They came to be described by many people over centuries as God worked in their lives.

Measurements and Experiments — Galileo Galilei

The Italian scientist Galileo Galilei (1564–1642) was the first to experimentally study the **acceleration of gravity** (the rate that an object speeds up as it is falling). He did not have a timing device, so he used his pulse rate (which did vary). When he dropped heavier and lighter objects of the same shape, they fell at the same rate.

After Huygens developed lenses, Galileo put them together and made a refracting telescope. Refracting refers to using lenses to bend light rays. It was assumed that objects outside of earth were perfectly shaped because they were closer to heaven. With his telescope, Galileo saw craters on the **moon**, four moons around Jupiter, and sun spots. He saw

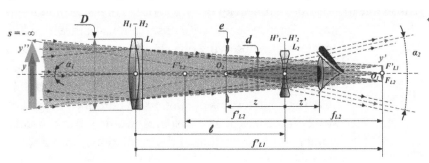

Galilean telescope

phases of Venus (like the phases of the moon) indicating that Venus orbited around the sun. These sound like great discoveries, but they went against the teachings of the Catholic Church at that time. The doctrine of the Church was that earth was the center of God's attention; therefore, it was at the center of the universe, and the planets and the sun orbited around earth. Galileo was declared a heretic and judged by the Inquisition court. The penalty was death, but because of the persuasion of close friends, he was placed under house arrest. Before Galileo, Copernicus supported the idea of planets orbiting the sun. This led to the idea of a conflict between the Church and science. A major problem was that these doctrines were not biblical doctrines, but ideas developed from philosophy. A safeguard is to use chapter and verse from the Bible where a doctrine is stated in clear terms, which has the authority of God in Scripture.

In 1638, Galileo published his life work in a book entitled *Discourses and Mathematical Demonstrations Concerning Two New Sciences*. It is usually referred to as *Two New Sciences*. The two sciences were the science of moving objects (velocity, acceleration, momentum, etc.) and the science of non-moving objects. It was smuggled out of Italy because of the conflict with the Church in Rome and published in Leiden, Netherlands. This was considered to be the first modern scientific textbook because it described the universe using laws that could be understood by the human mind. Because of the conflict with the Church at that time and the idea that humans were able to understand the world around them, science became in the minds of many to be the highest source of knowledge. Many felt that they could understand all they needed to know without God. God has given us a tremendous ability to understand His creation because we are created in His image. There are many very intelligent people today who feel they have no need for God in their lives. This goes back to a choice to be independent of God. Galileo, however, believed in God and His role in creation and maintaining the universe. Because God has been so gracious to us by giving us the remarkable gift of intelligence, we still need to recognize our need for Him. You will meet many very intelligent people who reject Christ and worship science. Remember that their view is based upon a choice to exclude God and not upon evidence. I often wonder what science will be like in the new heavens and earth without sin and imperfection. Science is a gift from God, as Galileo recognized, to be used wisely, but our lives go even beyond that to appreciate the greater wisdom given by His Spirit through the Scriptures. Perhaps God can use you to reach many in the sciences to lead them to redemption in Christ. With the Holy Spirit, we can understand and enjoy physics much more.

Science has been a valuable gift from God. He created the natural laws and gave us the ability to use them to design many blessings such as our computers, modern medicine, automobiles, airplanes, etc. The greater gift He gave us was His Son whom He demonstrated as authentic by the miracles that Christ performed as was prophesied centuries before by the prophets. The miracles performed by the Apostles as well clearly demonstrated God's sovereignty over the natural laws that He created.

Galileo used what became known as the **scientific method**. This involves using experimentation to test ideas rather than accepting what seems reasonable. It seemed reasonable to Aristotle that heavier objects should fall faster than lighter objects, but it just is not so. Science begins with observations from which ideas (**hypotheses**) are developed. In observing objects falling, Galileo noticed that heavier objects did not fall as fast as he thought they should. From that he predicted that the heavier and lighter objects would fall at the same rate. Experiments had to be designed to test his prediction that dropped objects of the same shape (to avoid differences in air resistance) would fall at the same rate. His experiments supported his prediction. If the predictions did not work out as expected, the original idea would need to be altered or replaced. A well-supported hypothesis is called a **theory**.

To be able use the scientific method, experiments must be repeatable. Many consider science to be the most reliable way of discovering truth because it relies on repeatable experiments with measurements. It also involves peer review where others repeat the experiments to confirm the results. But a major limitation of science is that it is impossible to think of all possible explanations (hypotheses) for the observations, all the possible predictions from the hypotheses, and all the possible experiments to test the predictions. As new hypotheses, predictions, and experiments are devised, hypotheses and theories are always being replaced by better ones. When science is treated as the ultimate source of knowledge, the conclusion is that there are no absolutes (nothing can be known for certain). The critical point is that there is a greater source of knowledge — God and the Bible that He provided. The Bible must be handled correctly and not made to say things that it does not say. Galileo recognized this when he said that "God is known by nature in His works, and by doctrine in His revealed word."[1]

It was an error to make the earth the center of the universe doctrine because the Bible does not state that. Our knowledge from science grows and is constantly changing, but the truths of the Bible are the same. God is the greatest authority. One of our greatest limitations is that it is hard for us to grasp the idea of the infinite, omnipotent, omnipresent loving God. Some today compare the opposition that Copernicus and Galileo experienced to the opposition of creation against evolution. This is not a valid conclusion because these are different

---

[1] "Letter to the Grand Duchess Christina of Tuscany, 1615"; *Discoveries and Opinions of Galileo: Including The Starry Messenger (1610), Letter to the Grand Duchess Christina (1615), and Excerpts from Letters on Sunspots (1613), The Assayer (1623).* Translated with an Introduction and Notes by Stillman Drake. NEW YORK: Anchor Books, 1957. Pg. 183. The quote is a shortened version of Tertullian's quote in the book *Adversus Marcionem*.

ideas, and creation is described in the Bible. Without the guidance of the Holy Spirit, it is impossible to realize that God's Word is greater than anything that we can propose. Over the years, I have noticed that as good, honest objective research is done in science, it does not contradict the Bible. The conflict lies in the opinions of some. It is true that the Bible is not a textbook of science, but where it overlaps with science it is true because it is based upon the integrity of God.

Measurements in Physics

Measurements are part of the wisdom of God that went into the original creation. Proverbs 16:11 states that "Honest weights and scales are the Lord's; All the weights in the bag are His work."

To test the hypothesis that heavier objects had the same acceleration of gravity as lighter objects, Galileo had to be able to measure the acceleration of gravity of many objects of differing weights.

Length is a fundamental measurement. In the English system, we use inches, feet, yards, and miles. In the metric system everything is divisible by 10, which is much easier to use, which is why it is used exclusively in science. The metric system was developed by Antoine Lavoisier. He was honored in France by being placed in charge of taxation right before the French Revolution. He was one of the first to lose his head to the guillotine. This does not mean that you will go to the guillotine if you use metrics.

The basic unit in the metric system for length is the meter.

The prefix **deci-** means $\frac{1}{10}$;

centi- means $\frac{1}{100}$;

milli- means $\frac{1}{1000}$ and

kilo- means $\times 1,000$.

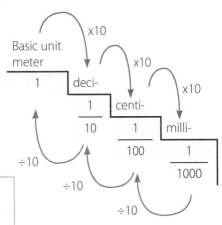

There are 10 decimeters in a meter; 100 centimeters in a meter; 1,000 millimeters in a meter and 1,000 meters in a kilometer.

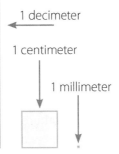

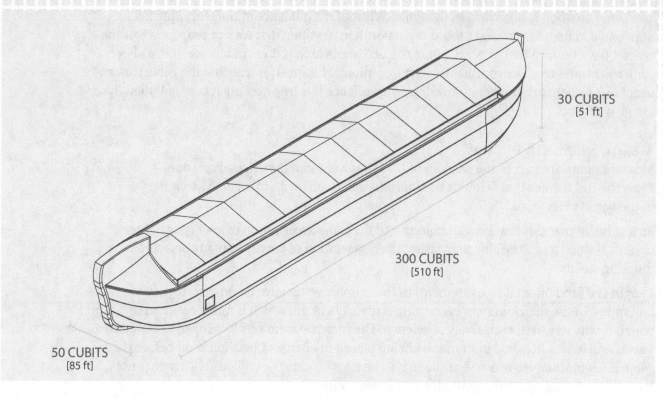

Noah had a clear understanding of measurements because God gave him specific instructions in building the Ark. The instructions were in cubits, which was the unit of length used at that time. There are 1.7 feet in a cubit and 0.308 meter in a foot.

The Ark was 50 cubits wide.

$$50 \text{ cubits} \times \frac{1.7 \text{ feet}}{\text{cubit}} = 85 \text{ feet}$$

$$85 \text{ feet} \times \frac{0.308 \text{ meter}}{\text{foot}} = 26.2 \text{ m wide (m stands for meter)}$$

The Ark was 30 cubits high.

$$30 \text{ cubits} \times \frac{1.7 \text{ feet}}{\text{cubit}} = 51 \text{ feet}$$

$$51 \text{ feet} \times \frac{0.308 \text{ m}}{\text{foot}} = 15.7 \text{ m high}$$

The Ark was 300 cubits long.

$$300 \text{ cubits} \times \frac{1.7 \text{ feet}}{\text{cubit}} = 510 \text{ feet}$$

$$510 \text{ feet} \times \frac{0.308 \text{ m}}{\text{foot}} = 157.1 \text{ m long}$$

So, the volume of the Ark was 26.2 m × 15.7 m × 157.1 m = 64,621.5 m³ (cubic meters), because you multiply length × width × height.

That was some boat!

If you get a chance, go see the Ark Encounter in Williamstown, Kentucky, which is a replica true to size of the ark built by Noah to the biblical dimensions.

Did you notice how cubits were converted to feet and meters? A cubit is 1.7 feet and 0.518 meter is the same length as a foot. In the expression cubits × $\frac{\text{feet}}{\text{cubit}}$, cubits are divided by cubits which is 1 leaving 1 × feet. Another way to state it is to say that the cubits cancel each other. Do not get bothered if you do not understand why this is so, pay attention to the process. The understanding can come later in a math course. Right now, we are using it as a tool. It is like knowing how to use a computer without knowing how to program it.

The basic unit of mass in the metric system is the **gram** (g) and the basic unit of time is the second (s).

Density is defined as the mass of something divided by the volume. The average density of sea water is $\frac{1.027 \text{ kg}}{\text{m}^3}$. To float, the Ark would need a density (ρ) just below that of sea water — so its density would have had to have been less than 1.027 $\frac{\text{kg}}{\text{m}^3}$. The volume ($V$) of the Ark would have been 64,621.5 m³. For its density to be less than $\frac{1.027 \text{ kg}}{\text{m}^3}$, its mass ($m$) would have to have been just under. . .

$$\rho = \frac{m}{V}$$

$$\frac{x \text{ kg}}{64,621.5 \text{ m}^3} = \frac{1.027 \text{ kg}}{\text{m}^3}$$

$$x \text{ kg} = \frac{1.027 \text{ kg}}{\text{m}^3} \times 64,621.5 \text{ m}^3 = 66,366 \text{ kg}$$

A kg is 1,000 grams so 66,366 kg = 66,366,000 grams!

We have no way of measuring the mass of the Ark with all of the animals and Noah and his family on board, but we can find it with a little math. This is a method used often in physics. You can calculate things from what you can measure. Its validity lies in that it is based on what is measured and using a valid calculation procedure.

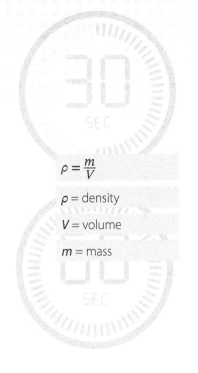

$\rho = \frac{m}{V}$

ρ = density

V = volume

m = mass

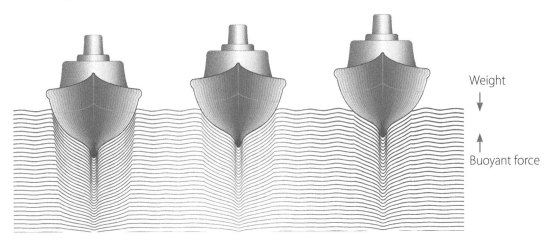

Weight

Buoyant force

If math is not your friend, do not panic because all the math that you need will be explained. It also makes more sense when you can see how it is applied.

Answer the questions on the related worksheet for this chapter in the Teacher Guide. You may look at the answers to correct your work or if you get stumped. You can write your answers on a separate sheet of paper if you wish to go back and redo them as a review as you prepare for the quiz on this chapter and the later exam. After you have reviewed this chapter and mastered the questions in the worksheet, then take the chapter quiz which will be graded by your teacher.

LABORATORY 1

Archimedes' Principle

REQUIRED MATERIALS

- Digital scale
- 12 cm × 12 cm piece of heavyweight aluminum foil
- Duct tape
- Metric ruler
- Package of at least 25 metal washers
- Fine tip permanent marker

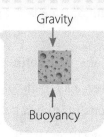

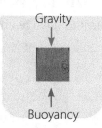

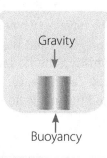

Introduction

Science is based upon observations, hypotheses, predictions to test the hypotheses, and determining whether the experiments support or reject the hypotheses. In this exercise, you will make some observations, make a hypothesis, and test it. Archimedes (287–212 B.C.) made observations like yours in this exercise. He noticed that some objects that were heavier floated while others that were lighter sank. Therefore, weight was not the only factor in determining whether something floated or not. Have you ever wondered why an aircraft carrier floats and a rock sinks? You have been making observations since you were born but probably have not wondered why they happened that way or tested them.

Purpose

This exercise provides experience in making observations and forming and testing a hypothesis.

observe

1. Take a 12 cm × 12 cm (centimeter) piece of aluminum foil and mark it using the ruler as shown in the diagram. (*Diagram L1.1*)

2. Cut the aluminum foil as shown in the diagram below and fold up the sides so that the triangle-shaped pieces on the corners overlap. With pieces of duct tape, tape the overlapping corners so that you have a square shaped "boat" that is 4 cm × 4 cm at the base with the sides 4 cm tall. (*Diagram L1.2*)

3. Find the mass of your boat using the digital scale. Find the mass of a washer. The mass of each are in grams.

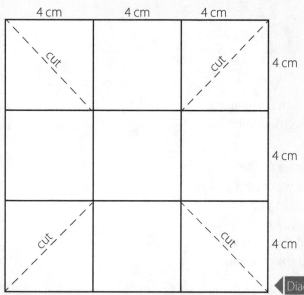

Diagram L1.1

12 | Physics

4. Mark the side of the aluminum boat 1 cm up from the bottom. (*Diagram L1.3*)

5. Place the aluminum boat in a body of water such as a bath tub or sink. Check it for leaks to be sure that no water enters the boat.

question

6. Carefully add washers to the boat until it sinks down to the 1 cm mark but is still floating. Count the number of washers that you added and multiply that number by the mass of the washer found in step 3. Add the mass (grams) of the washers to the mass of the boat for the total mass. Find the density of the boat and washers by dividing the mass (grams) of the boat and washers by the volume of the boat. The volume of the boat is 4 cm × 4 cm × 4 cm = 64 cm³.

research

7. Add additional washers to the boat until it sinks. Count the number of washers you added and multiply that number by the mass of the washer found in step 3. Add the mass (grams) of all the washers to the mass of the boat for the total mass. Find the density of the boat and washers after the boat sank. Divide the total mass of the boat and the washers (that made the boat sink) by its volume (64 cm³).

8. Record your data and observations in your lab report.

9. The density of water is $\frac{1 \text{ gram}}{\text{cm}^3}$.

hypothesis

10. Look at your data and observations and form an explanation (hypothesis) as to why the boat floated in step 6 and sank in step 7. Write out your hypothesis and how you came up with it. In other words, propose a reason why the boat floated in step 6 and sank in step 7.

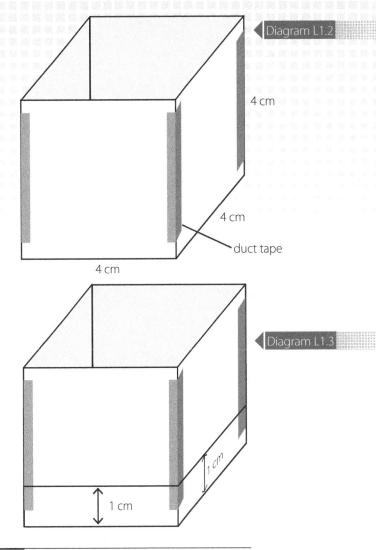

experiment

11. Now you need to test your hypothesis. A suggestion would be to repeat step 7 but add 1 less washer than in step 7 and observe whether it floats or sinks. Then find the density of the boat with its washers (as in step 7).

analyze

12. How would this support or reject your hypothesis? You can use your creative juices to come up with other ways to test your hypothesis.

conclusion

13. In your report, state how you tested your hypothesis and whether you supported it or rejected it and why.

CHAPTER TWO

SPEED, VELOCITY, AND ACCELERATION

OBJECTIVES

At the conclusion of this lesson the student should have an understanding of

- The branch of physics called mechanics
- Aristotle's concepts of moving and falling objects
- Galileo's concepts of moving and falling objects
- Galileo's testing ideas with experiments
- The application of graphs to the study of motion
- The concepts of displacement, average speed, instantaneous speed, average velocity, instantaneous velocity, and acceleration
- The use of scalars and vectors

Mechanics is the field of physics that deals with motion including speed, velocity, acceleration, force, Newton's three laws of motion, energy, and work.

Aristotle claimed that a force had to keep acting on an object to keep it in motion. Galileo's experiment with an inclined plane disproved this idea about force. Remember that Aristotle's ideas were accepted for almost 2,000 years. Galileo's work was bold going against what was widely accepted as common sense. A plane is a flat two-dimensional surface such as the surface of a rectangular board. Galileo placed two boards at an angle to each other shaped like a V. He placed a ball at the top of one board and allowed the ball to roll down the board and up the other board. Then the ball rolled back down and went back up the first board. It kept going back and forth between the boards until it gradually slowed down because of friction. (*Diagram 2.1*)

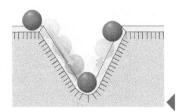

Diagram 2.1

He found that the same thing happened when he widened the angle between the boards. He widened the angle until the boards were almost straight. (*Diagram 2.2*) He reasoned that when the ball began rolling down a flat surface, it would keep going until another force (friction) slowed it down. Therefore, the ball did not require a constant force to keep it going and would keep going until something else slowed it down or stopped it. This idea was later expanded upon by Isaac Newton.

Some have said that Galileo dropped a heavier object and a lighter object off the Leaning Tower of Pisa and they both hit the ground at the same time, traveling at the same velocity. This was the opposite of what was taught by Aristotle. An important truth was learned by a simple experiment. An important principle in science is to test even what seems to be obvious. Aristotle's teachings were based on what he understood and what made sense to him. There are a lot of teachings that deny God's Word that seem reasonable to a lot of people, but that does not make them true.

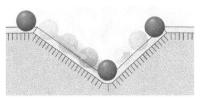

Diagram 2.2

Some very important concepts in the study of motion include displacement, speed, velocity, and acceleration.

Displacement is the movement of an object from one place to another — how far it has moved and in which direction. This can be illustrated with a graph.

Diagram 2.3

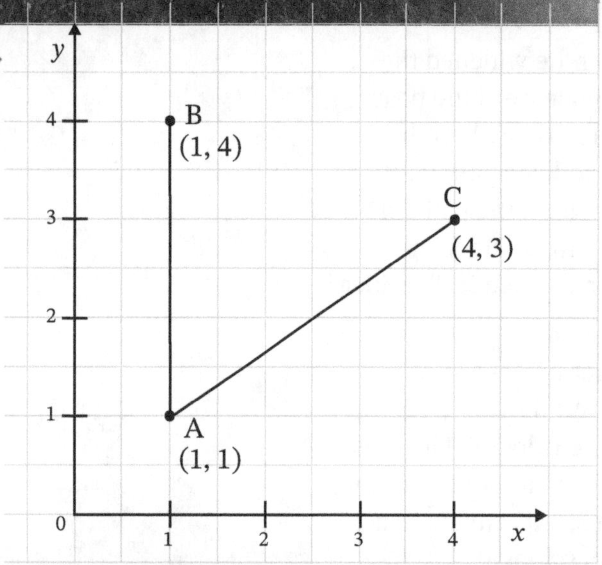

A graph has values of x (on the horizontal line) and y (on the vertical line). "0" is called the origin and both values of x and y are zero. Point A has the values $x = 1$ and $y = 1$. It is written as (1, 1) where x is the first value and y is the second value. Point B has $x = 1$ and $y = 4$. It is written as (1, 4). Point C is at position (4, 3). The displacement of an object from point A to point B is the length of the line from A to B. The displacement from point A to point C is the length of the line from A to C (*Diagram 2.3*). A simple way to measure the displacement is to measure the length of the line with a ruler.

The words speed and velocity are usually used to mean the same thing in everyday use. In physics, however, they have different meanings. Speed is how fast something is moving — 5 miles per hour read on a car's speedometer means that the car is going 5 miles in 1 hour. The speedometer does not tell you which direction the car is going. The speed limit is called that because it applies whichever direction you are going. If you travel 476 miles in 7 hours, your average speed is 68 miles per hour $\left(\frac{476}{7} = 68\right)$.

Speed

16 | Physics

This does not mean that you traveled 68 miles per hour for all 7 hours. Sometimes you went faster and sometimes slower. If you were clocked at exactly 69.5 miles per hour at exactly 2 hours after you started, this would be your instantaneous speed at the 2-hour mark.

The word velocity includes speed and direction. If you are going 68 miles per hour toward the west, it is your velocity. Just like speed, there is an average velocity and an instantaneous velocity.

Quantities like speed that do not have direction are called **scalars**. An example is 68 miles per hour. When quantity and direction are both given, they are called **vectors** such as 68 miles per hour toward the west.

Displacements are often described using graphs. Because displacement and velocity are both vectors (they both include direction) they can both be described using graphs. Average velocity is displacement divided by time. Time is on the x axis and displacement is on the y axis.

This graph shows an object going 3 meters (y axis) in 5 seconds (x axis). To be a vector, direction must be given — this object is going toward the east. The average velocity is displacement divided by time which is $\frac{3 \text{ meters}}{5 \text{ seconds}} = \frac{0.6 \text{ m}}{\text{s}}$. In physics, the units meters and seconds are usually used. The change in y (displacement) divided by the change in x (time) is called the slope. It is written as $\frac{\Delta y}{\Delta x}$ where the capital Greek letter delta Δ means "change in." $\frac{\Delta y}{\Delta x} = \text{slope} = \frac{\text{m}}{\text{s}}$, which in this case is the average velocity (Diagram 2.4).

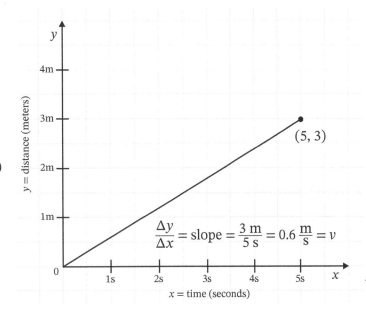
Diagram 2.4

At this point, it may seem like this is being made more complicated than is necessary. But as you will see, being able to make measurements and put them on a graph that shows the relationships between the measurements makes them easier to understand and come up with new ideas.

One more concept that needs to be considered is **acceleration**. Acceleration is changing velocity. Changing velocity may also include changing direction because direction is part of velocity. If a car goes around a corner at constant 10 miles per hour, it is accelerating because it is changing direction.

Just as velocity is the change in distance (displacement) divided by the change in time, acceleration is the change in velocity divided by the change in time.

$$a = \frac{\Delta v}{\Delta t}$$

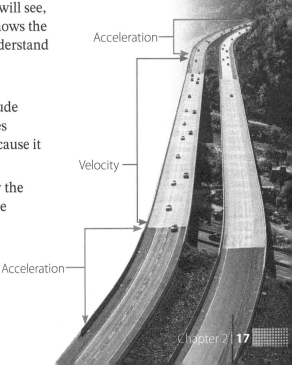

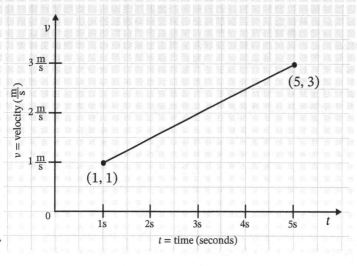

Diagram 2.5

This graph (*Diagram 2.5*) shows the velocity going from $1\frac{m}{s}$ to $3\frac{m}{s}$ from second 1 to second 5. Notice that the object started at $1\frac{m}{s}$ rather than from $0\frac{m}{s}$. The time went from 1 second to 5 seconds. This is where a graph comes in handy to show this.

$$\Delta v = v_f - v_0 = 3\frac{m}{s} - 1\frac{m}{s} = 2\frac{m}{s}$$

v_f is the final velocity and v_0 is the initial (beginning) velocity.

$$\Delta t = t_f - t_0 = 5\,s - 1\,s = 4\,s$$

t_f is the final time and t_0 is the initial time.

$$a = \frac{\Delta v}{\Delta t} = \frac{2\frac{m}{s}}{4s} = 0.5\frac{m}{s^2}$$

The units $\frac{m}{s^2}$ may look very odd. It means that an object had increased velocity ($\frac{m}{s}$) every second — $\frac{m}{s}$ divided by s is $\frac{m}{s^2}$ because s^2 is s × s. It looks like this. At 1 second the object's velocity is $1\frac{m}{s}$; at 2 seconds its velocity is $1.5\frac{m}{s}$; at 3 seconds its velocity is $2\frac{m}{s}$; at 4 seconds its velocity is $2.5\frac{m}{s}$ and at 5 seconds its velocity is $3\frac{m}{s}$. So, every second its velocity is increasing by $0.5\frac{m}{s}$.

Galileo reasoned that a falling object accelerated going faster and faster. Today we know that a falling object (without air resistance) accelerates at $g = 9.8\frac{m}{s^2}$. The letter *g* is used for the acceleration caused by gravity. When an object is falling, the average velocity is $9.8\frac{m}{s^2}$ for the first second; the average velocity for the second second is 19.6 (2 × 9.8) $\frac{m}{s^2}$ and the average velocity for the third second is 29.4 (3 × 9.8) $\frac{m}{s^2}$.

If air resistance is great enough that the object no longer accelerates while falling it is in **free fall**.

If you stood on a scale in an elevator, your weight would appear to increase as the elevator went up. When the elevator is not moving, the floor of the elevator is pushing up on you with the same force that gravity is pulling you down — that is why you are not moving. But when the elevator goes up, the floor of the elevator pushes on you with a greater force than gravity is pulling you down, so you go up and your weight on the scale appears to increase. If the elevator goes down at the same rate as free fall, you will record zero weight on the scale, and you will feel weightless.

When a rocket blasts off and goes into orbit around the earth it must achieve **escape velocity**. The acceleration of the rocket must exceed $9.8 \frac{m}{s^2}$ in the opposite direction as gravity. As the rocket gets farther from earth, g decreases. When the rocket gets twice as far from the center of the earth, g is reduced by $\frac{1}{4}$. Isaac Newton called this the inverse square law, meaning that it decreases by the square of the distance away from earth. The universal law of gravity is $F = \frac{G \times m_1 \times m_2}{d^2}$. As the distance between the objects increase, the force of gravity between them decreases by the square of that distance. The word *inverse* means that the force of gravity between the objects decreases while the distance between them increases. When the rocket gets 3 times farther from earth, the force of gravity decreases by $\frac{1}{9}$; g also decreases the same way because g and the force of gravity are proportionate to each other as $F = mg$. To stay in orbit around the earth, the rocket would have to adjust from vertical motion to horizontal motion. To keep from falling back to earth, the rocket would have to have a velocity of $11 \frac{km}{s}$.

LABORATORY 2

Velocity And Acceleration

REQUIRED MATERIALS
- Meter stick
- 2 small objects of different masses
- Digital scale
- Stopwatch or smart phone with a stopwatch app
- **Note:** You will need a partner to help with this lab.

Introduction

Mechanics is the study of motion which involves speed, velocity, and acceleration. Speed and velocity are defined as distance divided by time. For example, when an object moves 5 meters in 10 seconds, its average speed is $\frac{5 \text{ m}}{10 \text{ s}}$ or $0.5 \frac{\text{m}}{\text{s}}$. If you know the direction of movement, it is velocity instead of speed. Speed is called a **scalar** because it is only a quantity, but you do not include the direction of movement. Velocity, however, includes direction, so it is called a **vector**. This is important and gives you more tools to work with as you will see in the next few chapters. Acceleration is a change in velocity that can include changing the quantity of $\frac{\text{m}}{\text{s}}$ and change in direction, because velocity includes direction or both.

Contrary to Aristotle, Galileo found that objects of very different masses fell at the same rate. The acceleration of gravity does not depend upon the mass of an elephant. If you jump off a high cliff and an elephant jumps from the same spot after you, you will stay ahead of the elephant until you land. You will both fall at the same rate. Perhaps you will have just enough time to get out of the way before the elephant lands right after you. The acceleration of gravity means that falling objects keep increasing in their velocity as they are falling. All objects increase their velocity at the same rate unless they are affected by air resistance. Perhaps to give you a little more time to get out of the way, you could give the elephant a parachute when he jumps after you.

Purpose

The purpose of this exercise is to measure the velocity of a moving object, describe its acceleration, and demonstrate that objects of differing mass have the same acceleration of gravity.

Procedure

 observe

1. Measure and mark 2 meters using the meter stick. If you are using a hardwood or tile floor in the house be sure you are able to remove your marks. Take a ball and roll it between the marks a few times for practice.

2. Now roll the ball 3 times and time how long it takes the ball to roll from one mark to the other. Also, carefully observe the acceleration of the ball (speeding up, slowing down, and changing direction). Write these results in your lab report. You will need some help because it is too much for one person to keep track of.

? question

3. Take the average of the 3 times and record this value in your lab report. Is this speed or velocity? Note that the ball is rolling in each direction.

research

4. Take 2 different objects and measure their masses using a digital scale. Record these values. The scale gives you grams, but you need to convert these values into kilograms because that is the common unit of mass used in physics. Take the grams and divide by 1,000 to get kilograms. For example, $\frac{43 \text{ grams}}{1,000} = 0.043$ kilograms (or $\frac{43 \text{ g}}{1,000} = 0.043$ kg).

hypothesis

5. Drop the objects separately from at least a meter high — the higher the better. Measure the height with the meter stick. The tricky part here is to time their fall. Drop each object separately and measure from the time of release to when the object lands. Practice this a few times and then drop and time each of the objects at least 3 times and take the average for each.

experiment

6. Divide the distance the objects fell by the average time for each. This gives the average velocity of fall for each. *Caution:* Keep in mind that depending on the objects chosen, damage could occur on surfaces – so be careful in choosing where to do this part of the lab.

analyze

7. Compare the rate of fall for the 2 objects. If one object has twice the mass of the other and fell twice as fast, then you could say that the mass of the object affected the rate of fall. But if their rates of fall differ by much less than their masses differ, then you could account for the difference in the rates of falling to experimental error.

conclusion

8. Which measurement is the most subject to error? Why? This measurement is called the **limiting factor** of the procedure. So, who was right, Aristotle or Galileo?

CHAPTER THREE

FORCE & NEWTON'S 3 LAWS OF MOTION

OBJECTIVES

At the conclusion of this lesson the student should have an understanding of

- Concepts of inertia and force
- Motion of the moon
- Newton's first law of motion
- Newton's second law of motion
- Pound Newton conversions
- Newton's third law of motion

Sir Isaac Newton's own first edition copy of his book.

Galileo realized that once a force was applied to an object it would move and continue to move unless another force slowed it down or stopped it. Aristotle said that if you push an object across the floor, it will not keep moving if you stop pushing it. But the opposing force is friction between the object and the floor, which Aristotle did not consider. Galileo reasoned that if there was no friction between the floor and the object, the object would keep moving.

Isaac Newton was born the year Galileo died (1642). In this case, as is always true in science, everyone's work is built upon the work of others who went before them. This is no exception. Isaac Newton's work was built upon the work of Galileo. Your learning is based, as well, upon what others have learned before you. As we study and do research, we can create new ideas (hypotheses) to better explain our observations. God created matter out of nothing (*ex nihilo*) and because we are created in God's image, we can create new ideas.

Newton had developed his ideas of the laws of motion by 1665 but did not publish them until 1687 in a large treatise titled *The Mathematical Principles of Natural Philosophy,* or *Principia* as it is better known today. Remember that natural laws are observations that are always the same everywhere — like the Law of Gravity. Later it was realized that Newton's Three Laws of Motion were consistent except for objects smaller than atoms and at speeds close to the speed of light. We accept these laws because we see them operating the same everywhere — even on the moon and Mars.

Most people conclude that God is only present and needed when miracles (exceptions to natural laws) are performed. This leads to a secular view of science where God is not part of the study of science. Natural laws are often called the nature of matter. In fact, even in this study, you may have wondered why God is brought up when this is supposed to be a science book. But this is assumed (a conclusion by choice — not tested). Newton looked for order and for laws in the world around him because he believed that God was wise and did things in an orderly fashion; therefore, consistent laws should be part of His creation. He expected to see order around him and found it. He did not expect to see disorder coming from the mind of God. Some have suggested that these laws were not observed before because those coming from pagan religions believed in gods that were not orderly and consistent.

Around 1665, the plague broke out in England which took many lives. Newton left Oxford University and returned home to the countryside to escape the plague, which gave him time to organize his thoughts and develop new ideas. In working through the laws of motion, he needed to develop the tools of mathematics to better explain his observations. He called these new tools *calculus*. He could calculate average velocities, but he did not have any means to determine the velocity of an object at any point in time. He could calculate the distance and time it took him to get home from Oxford, but he could not determine his velocity an hour after he left. This is where he used calculus. For many, the word *calculus* brings to mind a very difficult thing to learn. Rather, it is a matter of following a set of procedures that Newton so graciously already worked out for us.

Woolsthorpe Manor, Lincolnshire, England, birthplace of Sir Isaac Newton.

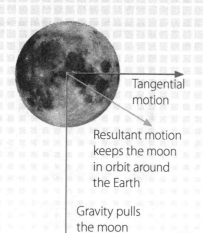

Tangential motion

Resultant motion keeps the moon in orbit around the Earth

Gravity pulls the moon toward Earth

Diagram 3.1

Legend has it that he saw an apple fall from a tree. Some say that it hit him on the head as he was trying to take a nap in the shade of the apple tree on a nice warm, summer afternoon. He concluded that gravity (from the ideas of Galileo) pulled the apple to the earth. Then he thought, why does not gravity pull the moon to the earth? After much thought, he realized that gravity does pull the moon to the earth, but that at creation God also applied a horizontal force to the moon so that when both forces act on the moon at the same time, the moon keeps going around the earth in a precise predictable pattern (*Diagram 3.1*).

Newton's first law of motion is that objects at rest tend to stay at rest and objects in motion tend to stay in motion. He noticed that it is more difficult to start moving an object of greater mass than one of less mass. It is likewise more difficult to change the velocity of an object of greater mass than one of less mass. Which would you rather move or stop — a jumbo jet or a shopping cart? The property of matter that resists moving or stopping is called **inertia**. The **mass** of an object is a measure of its inertia. The metric unit of mass is the kilogram (kg). A 50 kg object is harder to move than a 5 kg object and a moving 50 kg object is harder to stop than a 5 kg object.

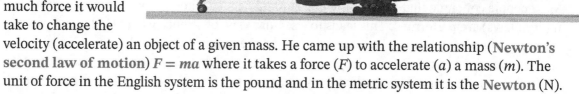

Newton followed this by wondering how much force it would take to change the velocity (accelerate) an object of a given mass. He came up with the relationship (**Newton's second law of motion**) $F = ma$ where it takes a force (F) to accelerate (a) a mass (m). The unit of force in the English system is the pound and in the metric system it is the **Newton** (N).

To accelerate a mass of 5 kg by $3 \frac{m}{s^2}$, the necessary force is 15 N.

1 lb = 4.4 N

$$F = ma = (5 \text{ kg})\left(3 \frac{m}{s^2}\right) = 15 \frac{\text{kg m}}{s^2} \text{ or } 15 \text{ N}$$

Notice that the units of a Newton is $\frac{\text{kg m}}{s^2}$.

1 pound is the same force as 4.4 N so . . .

$$\frac{15 \text{ N}}{4.4 \frac{\text{N}}{\text{pound}}} = 3.4 \text{ pounds}$$

Notice that in the above equation N is divided by N which cancels it out leaving the answer in pounds.

If a stationary 5 kg object is pushed so that it has an average velocity of $3 \frac{m}{s}$ for the first second and an average velocity of $6 \frac{m}{s}$ for the next second (an increase of $3 \frac{m}{s}$ every second), it would have to be pushed with a force of 15 N, which is 3.4 pounds.

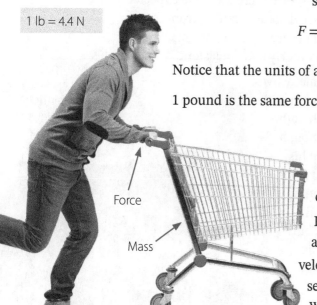

Force

Mass

Acceleration

If you weigh 130 pounds, your weight in the metric system is . . .

$$(130 \text{ pounds})\left(4.4 \frac{N}{\text{pound}}\right) = 572 \text{ N}$$

Your mass would be . . .

$$F = ma \text{ and } m = \frac{F}{a} = \frac{572 \text{ N}}{9.8 \frac{m}{s^2}} = 58.4 \text{ kg}$$

Here is a practical application. The acceleration of gravity in the metric system is $9.8 \frac{m}{s^2}$. In the English system, the acceleration of gravity is $32 \frac{feet}{s^2}$.

The force of gravity (weight) on a 20 kg object is . . .

$$(20 \text{ kg})\left(9.8 \frac{m}{s^2}\right) = 196 \text{ N}$$

The force of gravity upon an object is its **weight**.

$$\frac{196 \text{ N}}{4.4 \frac{N}{\text{pound}}} = 44.5 \text{ pounds}$$

These results are consistent with Galileo's observation that the mass of an object does not affect its rate of fall because all objects accelerate at the same rate when they fall ($9.8 \frac{m}{s^2}$).

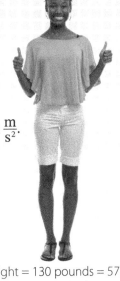

Weight = 130 pounds = 572 N
Mass = 58.4 kg

Acceleration of gravity = $9.8 \frac{m}{s^2}$

Acceleration of gravity = $32 \frac{feet}{s^2}$

If an object has more mass, it takes more force to accelerate it — it has more inertia to overcome. Therefore, the force of gravity (weight) is greater for more massive objects, but the acceleration of gravity is the same. Therefore, it can be said that mass is a measure of the inertia of an object.

It takes force to speed up, slow down, or change the direction (accelerate) of an object.

If a force of 10 N is applied to a 5 kg object what is its acceleration?

$$F = ma$$

$$a = \frac{F}{m} \text{ (divide both sides by } m\text{)} = \frac{10 \text{ N}}{5 \text{ kg}} = 2 \frac{m}{s^2}$$

Gravity is a force that the mass of an object exerts upon another object. The moon has a mass of about $\frac{1}{6}$ that of earth. If you weigh 160 pounds on earth, how much would you weigh on the moon? With a mass of $\frac{1}{6}$ that of earth, the moon can only exert $\frac{1}{6}$ as much gravity, so you would weigh . . .

$$\frac{1}{6} \times 160 \text{ pounds} = 27 \text{ pounds}$$

This is not a great weight loss program because your mass would still be the same. By the way, the acceleration of gravity is also different on the moon.

Moon

If you were in the International Space Station, you would be far enough away from earth's and the moon's gravitational pull that you would be weightless. This obviously does not mean that you would no longer exist, because your mass would still be same as on earth.

If you weighed 160 pounds, what would be your mass? The English unit of mass is the slug — not the slimy mollusk. First convert the pounds to Newtons.

$$(160 \text{ pounds})\left(4.4 \frac{N}{\text{pound}}\right) = 704 \text{ N}$$

International Space Station

If you have trouble following the math used here, just follow the procedure and the explanations will make more sense in math courses you take in the future.

$$F = mg \text{ (}g\text{ is the acceleration of gravity)}$$

$$m = \frac{F}{g} \text{ (divide both sides by }g\text{)} = \frac{704 \text{ N}}{9.8 \frac{m}{s^2}} = 71.8 \text{ kg}$$

Newton's three laws of motion describe the motion of objects. The first law describes how an object at rest tends to stay at rest. This means that a force must be applied to get an object moving and the second part of that law is that an object in motion will stay in motion unless opposed by another force.

A force is defined as that which causes a change in the velocity (speed up, slow down, and change direction) of an object. The second law describes how much force is needed to change the velocity of an object. So, the force needed to change the velocity of an object is the mass of the object times the change in velocity (acceleration).

This brings us to Newton's third law of motion. But first take a short break, get up and walk around the room. Carefully watch your feet. As you walk forward, in which direction are your feet pushing against the floor? Logic tells us that to move an object in a certain direction you must apply a force in that direction. If you push a small cart to the right, is it going to go to the left? Of course not. What direction are you pushing against the floor? Are you moving in the direction that you are pushing? Experience tells you that you must push the object you want to move. If you want to move a cart in front of you, you do not push against yourself. When you walk you are pushing against the floor — not yourself. You are pushing the floor (not you) in the direction opposite to the direction you are going. So, you are pushing a different object in the opposite direction. Did you ever imagine that walking could be so complicated? **Newton's third law of motion** is that for every force there is an equal and opposite force. In other words, your feet pushed against the floor in the direction opposite to the direction you were walking. The floor pushed back in the direction you were moving to move you across the room. It is silly to think that the floor decides it better obey Newton, so it gives you a push. We normally think of a force as something that we do to move something. But in this case, the floor is solid. It is not going anywhere soon hopefully. So, instead of the floor moving when you push on it, you move. Because you move, we say that a force was applied to you from Newton's first law of motion. Whenever you see a change in velocity of an object, you look for the cause — a force.

Stand close to a wall facing the wall. Push against the wall. Does the wall move? Hopefully not. Do you move, and in what direction? You move away from the wall in the direction opposite to the force you applied to the wall. The opposite and equal force is what pushed you away from the wall.

If you were in a small sail boat and had an electric fan and turned on the fan so that it blew into the sail, you would go nowhere. If a breeze picked up and blew into the sail, the boat would move. If you turned the fan around so that it blew out the back of the boat, the boat would move. What is the difference? The fan is in the boat, so it is part of the boat. Newton's third law applies when an object applies a force on a different object in the opposite direction to what you want to go. When you blow up a balloon and let it go, air rushes out of the hole in the balloon in the direction opposite to which the balloon flies around.

Force moves in the same direction as the boat so there is no movement

Bend over and grab your feet and lift yourself off of the floor. Obviously, this does not work. If you had someone else grab your feet and lift you, this could work if that person were strong enough and well balanced. What is the difference? The force has to be applied to a different object. Someone else must lift you. The same thing happened with the sailboat. The fan had to blow away from the boat. The amazing thing is that it works in the vacuum of space. This is how space craft are maneuvered. To have an equal and opposite force, there just must be an applied force. Remember that the laws of motion are observations of what happens. They are not explanations as to why they happen. The best explanation is that God in His great wisdom knew that these would be great for the function of the creation, so He decreed them and enforces them.

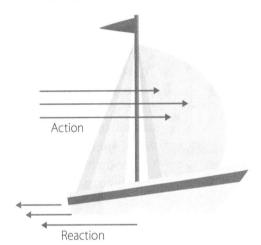
Force causes the boat to move forward in reaction to the wind

Through the remainder of this course, you will see these and other laws operate to make the universe function like a well-designed machine. Who said that physics is secular (without God)? By the way, no one has come up with a better reason why these laws should work and be so wisely designed, precise, and consistent.

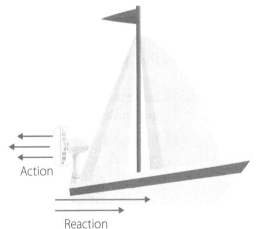
Force causes the boat to move forward in reaction to the fan

LABORATORY 3

Circular Motion And Newton's Laws Of Motion

REQUIRED MATERIALS
- Small heavy object
- Balloon
- Ball
- Bucket one-third full of water
- Book on a table

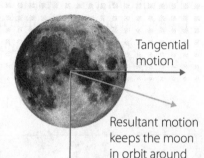
Tangential motion

Resultant motion keeps the moon in orbit around the earth.

Introduction
Isaac Newton realized that the moon's **rotation** about earth was the result of two forces acting upon it in different directions. He recognized one of the forces as gravity that pulled the moon toward earth. He reasoned that there had to be more to it than that because the moon has not crashed into earth. To give it the observed motion there had to be another force on the moon called a tangential force. This is at a right angle to a line pointing to the center of the earth.

You will demonstrate this by swinging a ball in a circular motion.

Newton's first law of motion states that an object at rest will stay at rest unless there is a force causing it to move, and that an object in motion will stay in motion unless there is a force causing it to stop or change its motion. He called the property that caused an object to stay at rest or in motion **inertia**. He observed that objects of more mass had more inertia. In other words, objects of more mass are harder to move and stop.

Gravity pulls the moon toward Earth

Newton's second law of motion is that the force necessary to accelerate (change the velocity) an object is the mass of the object multiplied by its desired acceleration, $F = m \times a$.

Newton's third law of motion is that when an object exerts a force upon another object, the other object exerts an equal and opposite force back on the first object.

Purpose
The purpose of this exercise is to demonstrate the circular motion of an object as the combination of a center directed force and a tangential force and Newton's laws of motion.

Procedure

1. When you swing an object in a circle, you are throwing it away from you, and by hanging onto it, you are keeping it from flying off. Go outside and swing a ball in a circle and after a couple of circles, let it go. Watch out so that you do not hit something with the ball. Notice that the ball goes at a right angle (90°) to your arm. The motion of the ball away from you in a straight line is its **tangential motion**. Because the ball went in a straight line after you released it, you know that the circular motion was the combination of the force you exerted (toward you) hanging onto the ball and the tangential force. Describe the motion of the ball before and after you release it. Draw a diagram showing the forces acting on the ball and the resulting force.

2. Take a bucket about one-third full of water. Swing it in a circle over your head so that the water stays in the bucket. Do this outside. You can tell your friends that you can hold a bucket of water upside down over your head without getting wet. As it did with the ball, your hand is pulling the bucket toward you while at the same time you are throwing it away from you. This time do not let go of the bucket. Describe the motion of the bucket and the water. Draw a diagram showing the forces acting upon the bucket and the water. Notice that you are touching the bucket, so it behaves like the ball. But you are not touching the water, so what holds it up against the bottom of the bucket?

3. Blow up a balloon and let it go. Think through your answers to these questions and be sure that they make sense. What happened to the balloon? In which direction did the air escape from the balloon? What was the direction of the force that pushed the balloon as it flew? What pushed the balloon? Where did this force come from? How did the balloon obey Newton's third law of motion?

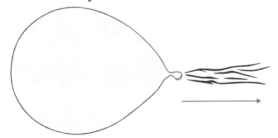

4. Hold your arm out in front of you with a heavy object in your hand. You can feel gravity pulling the object and your hand down. Are you pushing upward against gravity to keep your hand from dropping? Gravity and the force exerted by your muscles are equal and opposite forces. Which of Newton's laws is at work here? Explain.

5. Lay a book on a table. Push the book. How far does it move before stopping? Which of Newton's laws of motion are being observed here? Explain. Describe all the forces acting upon the book. You might want to review this in chapter 3. *Hint* – there are four.

CHAPTER FOUR

VECTORS

OBJECTIVES

At the conclusion of this lesson the student should have an understanding of

- A vector, resultant, vector diagrams
- Vectors used to find the resultant of going multiple distances in different directions
- Vectors used to find the resultant of applying multiple forces in different directions

A vector is one of the most useful tools in physics. It is a way of diagramming the quantity and direction of distances, forces, velocities, and accelerations. The direction is very important, as seen in this example. If you need to push a car to a service station and you and your friend are pushing on opposite ends of the car, you have a problem. If you both push on the back of the car together, you will be more successful. A vector is diagrammed as an arrow. The length of the arrow indicates the quantity of the vector and the direction of the arrow indicates the direction of the vector. If you push a box with a force of 10 pounds and your friend pushes on the box in the same direction with a force of 5 pounds, the total force on the box is 15 pounds as shown (*Diagram 4.1*).

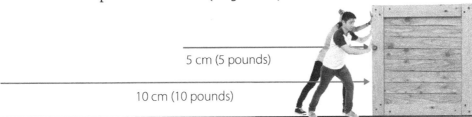

Diagram 4.1

The longer arrow is 10 cm (centimeters each of which is $\frac{1}{100}$ of a meter) long and the shorter arrow is 5 cm long. The 10 cm stands for 10 pounds and the 5 cm stands for 5 pounds.

In this example you and your friend are pushing on opposite ends of the box (*Diagram 4.2*).

Diagram 4.2

You are pushing with a force of 10 pounds to the right and your friend is pushing with a force of 5 pounds to the left. The total force is 5 pounds to the right. You can see that by subtracting the 5 pounds from the 10 pounds because they are in opposite directions. You can also diagram them this way (*Diagram 4.3*).

Diagram 4.3

The resulting force of several vectors together is called the **resultant**. When these terms are introduced, it is wise to learn their meanings right away. Often, something seems to appear very complicated because odd terms are used, and their meanings have not been learned. Keep this in mind throughout this study.

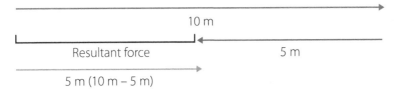

Diagram 4.4

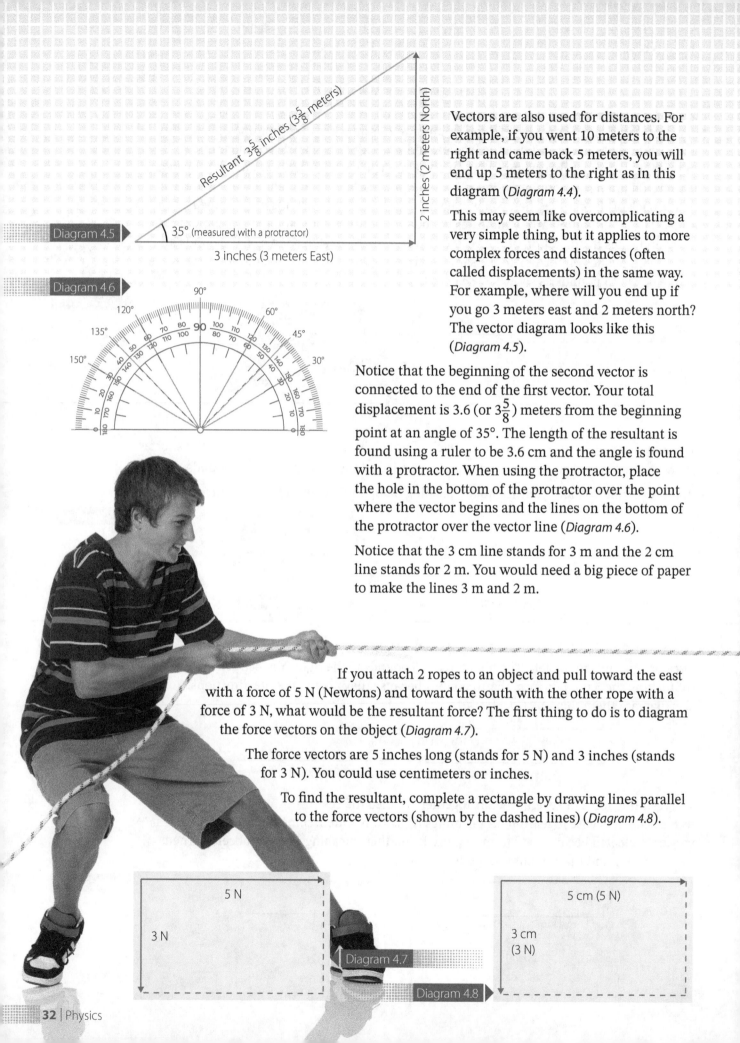

Vectors are also used for distances. For example, if you went 10 meters to the right and came back 5 meters, you will end up 5 meters to the right as in this diagram (*Diagram 4.4*).

This may seem like overcomplicating a very simple thing, but it applies to more complex forces and distances (often called displacements) in the same way. For example, where will you end up if you go 3 meters east and 2 meters north? The vector diagram looks like this (*Diagram 4.5*).

Notice that the beginning of the second vector is connected to the end of the first vector. Your total displacement is 3.6 (or $3\frac{5}{8}$) meters from the beginning point at an angle of 35°. The length of the resultant is found using a ruler to be 3.6 cm and the angle is found with a protractor. When using the protractor, place the hole in the bottom of the protractor over the point where the vector begins and the lines on the bottom of the protractor over the vector line (*Diagram 4.6*).

Notice that the 3 cm line stands for 3 m and the 2 cm line stands for 2 m. You would need a big piece of paper to make the lines 3 m and 2 m.

If you attach 2 ropes to an object and pull toward the east with a force of 5 N (Newtons) and toward the south with the other rope with a force of 3 N, what would be the resultant force? The first thing to do is to diagram the force vectors on the object (*Diagram 4.7*).

The force vectors are 5 inches long (stands for 5 N) and 3 inches (stands for 3 N). You could use centimeters or inches.

To find the resultant, complete a rectangle by drawing lines parallel to the force vectors (shown by the dashed lines) (*Diagram 4.8*).

32 | Physics

Then draw in a line that is a diagonal from the upper right to the lower left. This is the resultant force. Measure the length of the resultant line with a ruler. The number of centimeters is the resultant force in Newtons. With your protractor, measure the angle from the 5 N vector (*Diagram 4.9*).

The length of the resultant is 6.9 inches, which stands for 6.9 N, and the angle is 31° toward the southeast (SE).

If you travel 1.5 miles south, 4 miles east, and then 1 mile north, where will you end up in relation to your starting point?

The distance vectors are shown (*Diagram 4.10*).

The resultant vector is found by connecting the beginning of the first vector (1.5 cm) to the end of the third vector (1 cm) (*Diagram 4.11*).

With a ruler, the length of the resultant vector is 4.1 cm (stands for 4.1 miles), and with a protractor the resultant is SE at an angle of 83°.

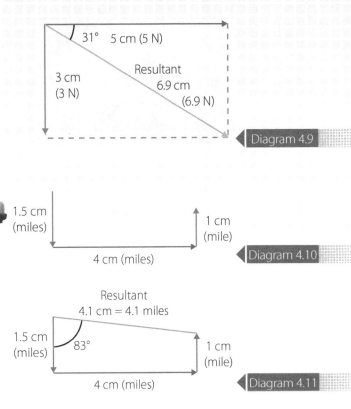

All these diagrams are 2-dimensional as drawn on a graph with the *x – y* axes. They can be drawn in 3 dimensions as well with a *z* axis that goes into and out of the page. They look like this (*Diagram 4.12*).

These are very helpful for determining displacements and forces for objects in motion — such as a space craft going to the moon.

A fourth dimension would be time. This considers where an object is at different points of time — such as a car 10 minutes after it leaves. In theoretical physics, many dimensions can be considered using vector matrices. If you are interested, this will come up later in advanced college studies.

Here is one more example. What would be the resultant force if you applied a force on an object of 3 N to the west, 1.5 N to the south, and 1 N to the east? (*Diagram 4.13*)

Draw the resultant vector as a line from the beginning of the first vector to the end of the third vector. Measure the length of the resultant vector in centimeters using a ruler. The number of centimeters is the number of Newtons (2.4 cm = 2.4 N). The angle from the beginning of the first vector to the end of the third vector is found with a protractor to be 36° (*Diagram 4.14*).

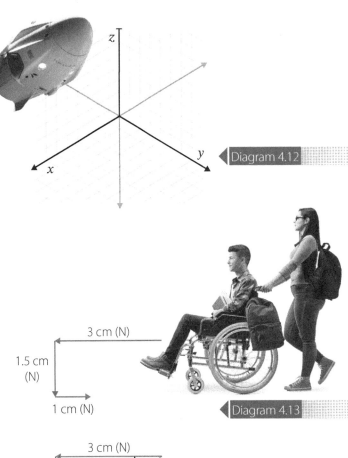

Chapter 4 | 33

LABORATORY 4

Vectors

REQUIRED MATERIALS

- A 6 inch (approximately) long section of 2×4 with an eye hook screwed into opposite ends
- 3 spring scales
- Protractor
- Metric ruler
- Meter stick
- String

Introduction
This chapter has demonstrated how the combined effects of several vectors can be determined. If an object is displaced in different lengths in different directions, its pathways can be diagrammed as vectors to determine where the object will end up. When several forces at different directions and magnitudes are imposed upon an object at the same time, their combined effect can be determined using vector diagrams. Since velocities and accelerations are also vectors, their combined effects can be determined using vector diagrams.

Purpose
The purpose of this laboratory exercise is to compare the results of vector diagrams with the resultants of physical forces and displacements.

Procedure

 observe

1. Connect a spring scale to one of the eye hooks on the piece of 2×4 and pull on it so that it moves along the surface. How many Newtons of force did you have to apply? You can determine the force from the spring scale as you pull on it. Describe what happens to the 2×4. Notice that you must overcome the friction between the 2×4 and the surface supporting it. (*Diagram L4.1*)

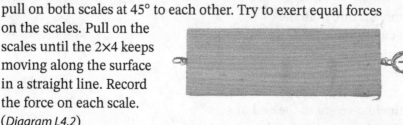

Diagram L4.1

 research

2. Connect 2 spring scales to the eye hook on one end of the 2×4 and pull on both scales at 45° to each other. Try to exert equal forces on the scales. Pull on the scales until the 2×4 keeps moving along the surface in a straight line. Record the force on each scale. (*Diagram L4.2*)

Diagram L4.2

34 | Physics

hypothesis

3. Use a vector diagram to determine the resultant of the two forces. Does the resultant equal the single force in procedure 1? If not, why? This answer comes from your reasoning.

 experiment

4. With the two scales connected to the eye hook, place one of them at an angle of 60° and the other at 30°. Connect a third spring scale to the eye hook on the opposite end of the 2×4. Pull on all 3 scales at the same time so that the 2×4 does not move. (*Diagram L4.3*) Record the forces on all 3 spring scales. Find the resultant of the two forces pulling on one side. Does the resultant equal the force pulled in the opposite direction? If not, why?

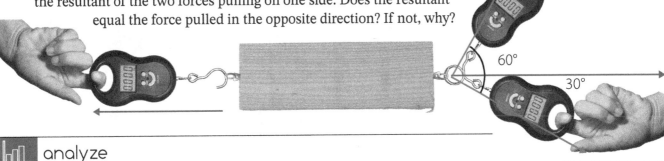

Diagram L4.3

 analyze

5. You will probably need to do this one outside. Use a meter stick to measure distances and mark your pathway with a piece of string. Walk 5 meters to the east, 3 meters to the north, and 2 meters to the west. Place a marker at your beginning point and your ending point. Measure the distance from where you started to where you ended up. Measure the angle between the resultant and your pathway from your starting point toward the east. (*Diagram L4.4*) Use the vector diagram below to find the resultant and angle.

conclusion

6. How do these values compare to what you measured? If they are not the same, why not? The reason could be quite simple.

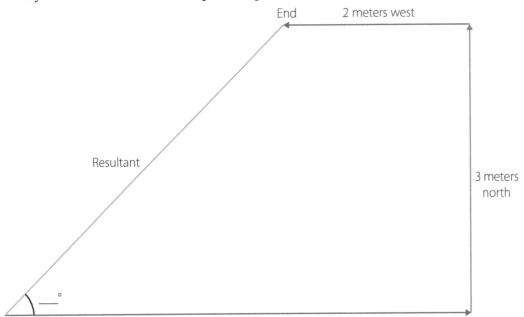

Diagram L4.4

CHAPTER FIVE

GRAVITY

OBJECTIVES

At the conclusion of this lesson the student should have an understanding of

- Fundamental forces of nature
- The concepts of gravity, mass, and weight
- The law of gravity and its application
- Weight and Newton's second law of motion
- Inverse square law
- Ocean tides
- Gravity force fields

The fundamental forces in nature include gravity, electromagnetic forces, and the strong and weak nuclear forces. These are "forces at a distance," meaning that they do not touch the objects they affect. If I dropped a book on the floor and wanted to read it, I would have to bend over and pick it up. What if, instead, I just looked at the book and drew it to me? That would be weird. But these four forces operate at a distance without something having to touch the objects they affect.

The strong and weak **nuclear forces** operate within the nucleus of an atom holding it together. Electromagnetic forces operate within and between atoms holding molecules together. Gravity holds objects made up of many molecules in their proper relationships with each other. Gravity holds you on earth, moons around planets, planets within the solar system, stars in galaxies, and galaxies in galaxy clusters. In contrast, electromagnetic forces attract and repel while gravity only attracts.

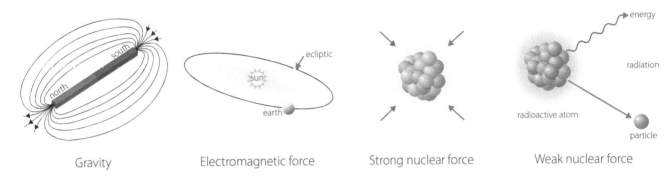

Gravity   Electromagnetic force   Strong nuclear force   Weak nuclear force

People usually think of these forces in one of three ways. One group chooses not to believe in God, so they view these forces as holding and maintaining the universe without a need for God. Another group acknowledges God but chooses to believe that God created these forces and then let the universe function on its own. A similar position is that God let these forces control the universe and only steps in when necessary. With this view, God only acts when there is a miracle. A third view is that these forces are the result of God's constant role in governing the universe. The third view is supported by Scripture. Colossians 1:17 states that "in Him [Jesus] all things hold together" (NIV). This means that these forces are the direct result of God constantly acting to produce and direct these forces. Hebrews 1:3 states that "He [Jesus] . . . upholds all things by the word of His power" (NASB). This means that God maintains the universe and is taking it to its ordained destiny. This means that He maintains these forces through time. Some would claim that this is unscientific, which is true. It is unscientific. Science is based upon what can be observed and verified with repeatable experiments. This would be the reaction without considering God. But knowledge can also be gained outside of science by direct revelation from God who created the universe and the forces that act within it. You will come across people from all these points of view. It is important that you understand that their views are based upon what they assume — not what they can prove. If you accept the authority of God's Word, you can support the third view. Why are these forces consistent throughout the universe and time? The answer is God's sustaining wisdom and power.

Normally, when we think of force we think of pushing or pulling something. If you were leaning against a wall, you would not expect the wall to push you away. Recall Newton's third law of motion. For every force there is an equal and opposite force. If you push against a wall, you will move away from the wall because of the equal and opposite force. Even in this case, the wall did not decide to push you away. The force originated from you when you pushed against the wall.

There is a case where an object does pull on another object. This is **gravity**. Earth pulls you toward its center. The force of this pull is called your **weight**. The greater your **mass**, the more mass gravity pulls, and your weight is greater. The sun pulls on earth, keeping it in orbit around the sun. Earth pulls on the moon keeping it in orbit around earth.

The pull of gravity is a two-way action. The sun pulls on earth and earth pulls on the sun. Gravity pulls objects toward each other. Newton demonstrated that this force can be expressed by the equation . . .

$$F = \frac{G \times m_1 \times m_2}{d^2}$$

which is called the universal law of gravity.

F is the force of gravity between two objects of masses m_1 and m_2. G is called the gravitational constant (meaning that it is always the same). It converts the units of the equation into Newtons (force). G is $6.670 \times 10^{-11} \text{ N} \frac{m^2}{kg^2}$. You do not need to memorize it. It will be given if you need it. Remember also that d is the distance between the **centers of the mass** of the two objects. The center of mass (also called **center of gravity**) is the point where all the mass of an object would be if it could be concentrated at one point. When an object rotates as it falls, it rotates around its center of mass.

If you did not have G in the equation, you would have $\frac{m_1 \times m_2}{d^2}$ which has the units of $\frac{kg \times kg}{m \times m}$ which is $\frac{kg^2}{m^2}$

This does not give you Newtons which are units of force. But the units of G are $\frac{N\,m^2}{kg^2}$. If you multiply $\frac{N\,m^2}{kg^2} \times \frac{kg^2}{m^2}$, the $\frac{m^2}{m^2} = 1$ and the $\frac{kg^2}{kg^2} = 1$ and you end up with N. The purpose of G is to make the units come out right. That is why the same value can be used every time. Here is an example of its use. The average distance between the earth and the moon is 3.84×10^8 m (that is 384,000,000 m). The exponent 8 means that the decimal is moved over to the right 8 places. The mass of earth is 5.973×10^{24} kg and the mass of the moon is 7.2×10^{22} kg. The equation becomes:

$$F = \frac{\left(6.67 \times 10^{-11} \frac{N\,m^2}{kg^2}\right)(5.973 \times 10^{24} \text{ kg})(7.2 \times 10^{22} \text{ kg})}{(3.84 \times 10^8 \text{ m})^2} =$$

Do the numbers first, then the exponents.

$$\frac{(6.67)(5.973)(7.2)}{(3.84)(3.84)} = 19.45$$

$$\frac{(10^{-11})(10^{24})(10^{22})}{(10^8)(10^8)} = 10^{19}$$

When you multiply exponents, you add them and when you divide exponents you subtract them, which is -11 + 24 + 22 - 8 - 8 = 19. This gives the answer as 19.45×10^{19} N or 1.945×10^{20} N (which is the preferred way of writing it). This is 194,500,000,000,000,000,000. Just what you have always wanted to know — the force of gravity between the earth and the moon. This also provides good practice using exponents and demonstrates how such distant things that cannot be directly measured can be found by using the right calculations.

Part of the beauty of physics is that you can find answers that you cannot directly measure. This unique ability to see mathematical relationships and come up with different ways to play with equations to find solutions to difficult problems is part of being created in the image of God. This displays our creative side. We cannot create matter out of nothing, but we can come up with answers to some things that are impossible to measure directly. The masses of earth and the moon and the average distance between them were calculated using their orbits and times that it takes the moon to orbit around the earth.

If you won a contest winning a round trip ticket (hopefully, round trip) to the moon, you could weigh yourself on the moon. To compare that to the calculation, F would be your weight on the moon, m_1 would be the mass of the moon and m_2 would be your mass; d would be the distance between your center of mass and the center of mass of the moon. The mass of the moon (m_1) is about $\frac{1}{6}$ the mass of earth, so on the moon you would weight about $\frac{1}{6}$ of what you would weigh on earth. If you weigh 120 pounds on earth, you would weigh $\frac{120}{6}$ or 20 pounds on the moon, though your mass would be the same.

Notice the term d in the equation. As you get farther from the center of the earth, d increases, and the force of gravity decreases. This is because you are dividing by d^2. If you went up in a space craft and doubled your distance from earth, d^2 would become $(2d)^2$ which is $4d^2$. That means that your weight would be reduced by $\frac{1}{4}$. You would be virtually weightless in orbit.

If you tripled your distance from the earth, your weight would be reduced by $\frac{1}{9}$. Some of my students that lived on a mountain weighed themselves before coming to class. After they got off the mountain, they weighed themselves again and they weighed a pound more. So, if you want to lose weight, live on a mountain. The diminished effect of gravity was called the **inverse square law** by Newton because it decreases by the square of the distance as you move away from its source.

In physics, there is often more than one way to find an answer. The method you use depends upon the information you have. If you know the masses of the two objects and the distance between their centers of mass, you would use the universal law of gravity equation. If you do not have that information but you do know the acceleration of gravity, you would use this other method. Recall that Newton's second law of motion is $F = ma$. If we use the acceleration of gravity (the acceleration of a falling object) where $g = 9.8\ \frac{m}{s^2}$ on the surface of earth, $F = ma$ becomes $F = mg$. If your mass is 46 kg, your weight is . . .

$$F = mg = (46\ kg)\left(9.8\ \frac{m}{s^2}\right) = 450.8\ N\ or\ 451\ N$$

1 pound is 4.4 N, so $\dfrac{451\ N}{4.4\ \frac{N}{pound}} = 102.5$ pounds.

If your mass is 46 kg, what is your weight on the moon? The acceleration of gravity on the surface of the moon is $1.62\ \frac{m}{s^2}$ and $F = mg = (46\ kg)\left(1.62\ \frac{m}{s^2}\right) = 74.5\ N$ which is $\dfrac{74.5\ N}{4.4\ \frac{N}{pound}} = 17$ pounds. This is almost $\frac{1}{6}$ of your weight on earth.

Another effect of gravity involves ocean tides. Think of the tides as the shifting of water from side to side on earth like what happens when you slide back and force in a bathtub sloshing water from side to side. The water is higher on one side of the tub than the other — then it moves to be higher on the other side.

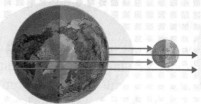

Gravity holds the oceans on the earth. But the moon and the sun also exert gravitational forces on the oceans. Even though the moon is much smaller than the sun, it is much closer, so it exerts a stronger force on the oceans. Again, we come back to Newton who realized that ocean tides are caused by the gravitational pull of the moon on opposite sides of the earth. The moon pulls the ocean water nearest it toward itself and pulls the other side of the earth toward it as well.

This produces a bulge of water on both sides of the earth. There is an important principle at work here. Gravitational forces are not shielded. The gravitational pull from the moon pulls on the opposite side of the earth as well. It is reduced by a greater distance from the moon, but not because any blocking activity of earth.

When the moon and sun are in line with each other, their gravitational forces act together and produce the greatest, highest tides — called **spring tides**. This has nothing to do with the season.

Spring tides

When the sun and moon are at right angles to each other, they do not add to each other as well, producing lower tides — called **neap tides**.

The positioning, sizes and distances of the sun and moon are signs of God's grace. If they were larger or closer, the tides would be very destructive. If they were smaller or farther away, the tides would be much smaller which would also be destructive. The tides are important for "flushing" (washing) the shorelines. They provide the shoreline wetlands that provide nurseries for young sea animals, nutrients for sea life, allow flood waters to spread out reducing potential damage, and detoxify many things being washed to the shores from rivers and streams.

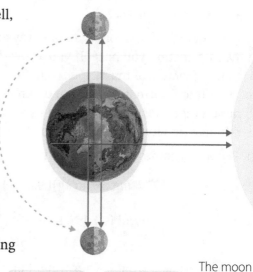
Neap tides

The moon exerts a greater pull because it is closer.

The previous discussion of patterns of spring and neap tides assumed that there was no land interfering with the movement of water. Did you catch that? There are land masses, so the water must move around continents and islands, producing a great deal of variety in the tides worldwide. This produces many very diverse marine habitats for many diverse life forms.

The phrase the moon interacts with the gravitational field of earth is used where gravity is thought of as a force field. The force field is the region over which gravity has its effect. The idea is that the space around earth is altered by gravity. How can you alter empty space? This means that an object experiences an effect when it enters that space. Force fields have magnitude and direction and can be shown with a vector diagram as shown by this one for earth's gravitational field. The lines pointing toward earth represent the force of gravity toward the center of the earth. Where the lines are farther apart the gravitational field is weaker, and where the lines are closer together (closer to earth) the gravitational field is stronger. The idea of a force field will come in handy later when considering electric and magnetic fields (*Diagram 5.1*).

Planets also have gravitational fields and pull on each other. The orbit of a planet around the sun is an ellipse (egg-shaped orbit). When their orbits come close to each other, perturbations are observed. This is where the planet's orbit is slightly distorted. In the early 1800s, unexplained perturbations were observed for the planet Uranus. A Frenchman, Urbain Leverrier, and an Englishman, J.C. Adams, both independently used Newton's Law of Universal Gravitation and determined that an unknown planet's gravitational field was affecting Uranus. They calculated where this planet should be and discovered Neptune. Other perturbations of Uranus led to the discovery of Pluto in 1930 at the Lowell Observatory in Arizona.

Gravity is an attractive force between two or more objects. The powerful ocean tides are produced by the gravitational forces between Earth and the Moon, as well as the sun. The greater the mass of an object, the greater is its gravitational attraction. However, distance is also a factor to consider, for the further apart objects are the less they attract each other. It is this force that holds planets in their orbits around stars and stars within their galaxies.

Diagram 5.1

Increasing gravitational force

High and low tides in Landes, France

LABORATORY 5

Gravity

REQUIRED MATERIALS

- String about 50 cm (centimeters) long
- Nail (not a finishing nail)
- 2 fishing weights (different weights)
- Digital scale
- Meter stick
- Phone with a stopwatch app or a stopwatch

Introduction

The consistency of gravity produces consistency in many other phenomena, including the swinging of a pendulum. Before the digital age, a pendulum was often used as a method of measuring time — such as in a grandfather clock. A weight (called a bob) swings on the end of a string or other object. The time it takes the bob to make a complete swing is the period of the pendulum.

In this laboratory exercise, you will be measuring the period of swing and the length of the string for a pendulum. In making measurements you need to be aware of significant figures in your data. Your answer cannot be more exact than your least accurate measuring device. For example, if you use different tools to measure the length of something you might get values of 12.3 cm, 12.42 cm, and 12 cm. The average of these values is 12.25 cm. The least accurate piece of data was 12 cm, which has 2 digits. Therefore, the measurement used must be rounded off to 2 digits, which gives 12 cm as the final answer. Keep this in mind as you do this lab exercise.

If you measure the density of aluminum, you can check yourself by looking up its accepted value. To see how close your measurement is to the true value, use what is called the **percent error**. You can calculate this value using this equation.

$$\% \text{ error} = \frac{\text{experimental value} - \text{true value}}{\text{true value}} \times 100$$

If you get a percent error of 10%, you are 10% off from the true value. This is a good test of your equipment and measurement skills. A smaller percentage is better. There will almost always be some percent error. Usually, the more accurate your equipment, the more expensive it will be. How close you are to the true value is the **accuracy**.

Many times, however, you do not know the value of a measurement before you do it. An example would be if you are measuring the temperature of the water in a creek flowing behind your house. You cannot look up that value somewhere. In this case, you want your measurements to be as close to each other as possible. This is called the **precision**. If you are shooting arrows at a target and they all hit very close to each other in a tree away from the target, your accuracy is very bad, but your precision is very good. To measure precision, use the **percent difference**.

$$\% \text{ difference} = \frac{\text{largest value} - \text{smallest value}}{\text{average value}} \times 100$$

In this lab, you are measuring the period of a pendulum, which you cannot look up anywhere. In this case, you need to evaluate your data using the percent difference.

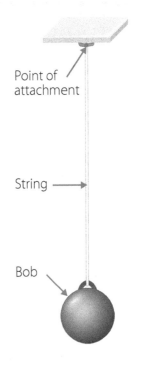

Point of attachment

String

Bob

Purpose
The purpose of this lab exercise is to . . .

1. Measure the period of a pendulum using bobs of different masses

2. Use the percent difference to evaluate your data

3. Observe the consistencies of a gravitational field upon different masses as shown by the consistency of the acceleration of gravity

Procedure

 observe

1. Pound a nail part way into an overhead structure. Get permission first, and be sure that you are not damaging something. This may even be a piece of 2×4 suspended a few feet off the ground.

question

2. Determine the mass of each of the fishing weights using the digital scale.

3. Attach the string to the nail with one of the fishing weights on the other end of the string. The length of the string from the nail to the fishing weight should be 40 cm.

4. Pull the weight 4 cm to the side and let it go so that it swings from side to side.

5. Using a stopwatch app on your cell phone (or a stopwatch), determine the time that it takes the bob to make 5 complete swings back and forth.

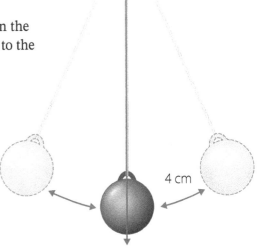

4 cm

6. Divide the time by 5 to get the period (T, time of 1 complete swing).

7. Repeat step 5 and step 6 5 times to get 5 values of T.

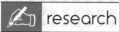

8. Calculate the percent difference for the 5 values of T. The smaller the percent difference, the better. Zero would be great.

9. Use this equation to calculate the value of g (acceleration of gravity) from the average of your 5 values of T.

$$g = \frac{4\pi^2 L}{T^2}$$

With your calculator, multiply $4 \times \pi \times \pi \times L$ (0.4 m) divided by $T \times T$. The units are $\frac{m}{s^2}$.

The lights on this ride at the Funfair in Rambouillet, France, show the movement of the massive pendulum.

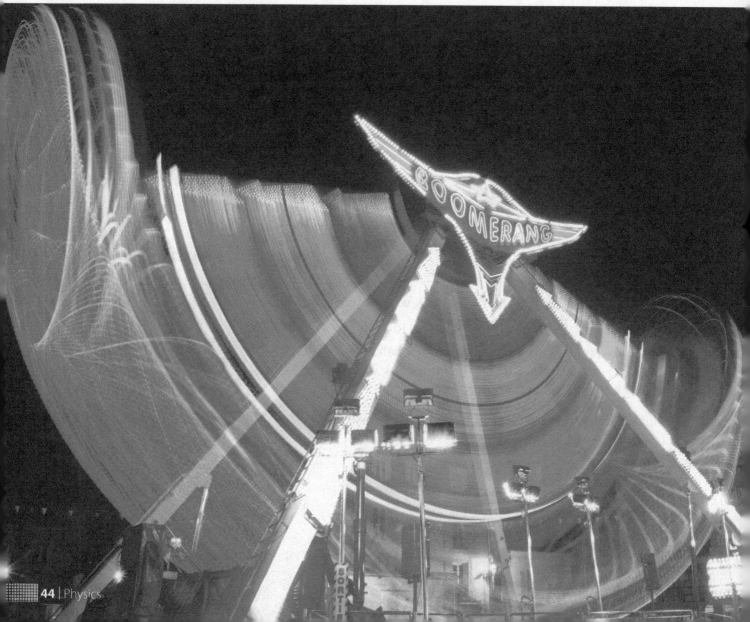

experiment

10. The value of g from step 9 is the experimental value. Be sure that it has no more digits than the measurement with the fewest number of digits — $9.8 \frac{m}{s^2}$ is the true value. Calculate the percent error from the equation given in the introduction. How did you do? The lower the percent error the better.

 analyze

11. Repeat steps 2 through 10 using the second fishing weight.

 conclusion

12. How do the experimental values of g compare using the two fishing weights? Calculate the percent difference for the 2 experimental values of g. The smaller the better. Small values support the consistency of g for different masses. A consistent value of g means that the gravitational field around earth is the same and different objects have different weights because of their different masses.

Metronomes help musicians keep their playing at a consistent tempo by providing a clicking noise to mark the pace of each musical interval.

The motion of a swing, going back and forth, higher and higher, demonstrates the swing of a pendulum and the force of gravity.

CHAPTER SIX

KINETIC ENERGY, POTENTIAL ENERGY, MOMENTUM, AND POWER

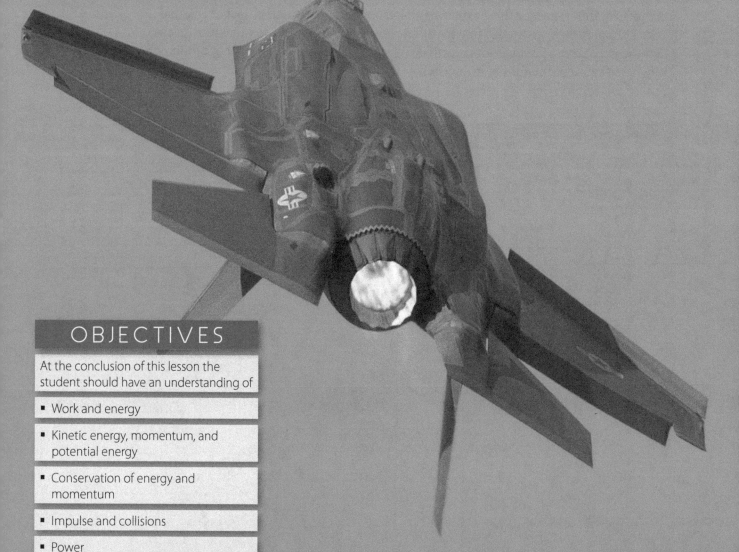

OBJECTIVES

At the conclusion of this lesson the student should have an understanding of

- Work and energy
- Kinetic energy, momentum, and potential energy
- Conservation of energy and momentum
- Impulse and collisions
- Power

Energy is usually defined as that which causes something to move. Does that mean that if I throw a ball with my hand, my hand is energy? No. it means that my hand has energy. We cannot see or taste energy. We can only see what it does. It is not an object even though the dictionary calls it a noun. It can also be defined as the ability to do work. We think of work as moving something. If I move my computer, then the work that I do is the force that it takes to move the computer times the distance that I move the computer. Or in physics terms, $W = F \times d$. The energy to move the computer is what supplies the force. if I am very tired, I say that I do not have much energy so I cannot move the computer. But if I had a good night's sleep and a hearty breakfast, I say that I have plenty of energy to move the computer. The units of work and energy are joules (J). Force is measured in Newtons (N) and distance is measured in meters (m), so work and energy are Nm (Newton meters). 1 Nm is also called a **joule** (J) (pronounced "jool") after the English physicist James Prescott Joule.

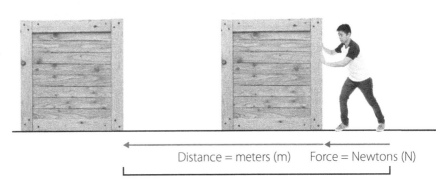

Distance = meters (m)   Force = Newtons (N)

Work = Newton meters (Nm) = joule (J)

Objects keep transferring energy to each other by bumping into each other. So where did the original energy first come from? When God created the heavens and the earth, the atoms and molecules and larger objects were set in motion. So we can say that God is the originator of energy that set things in motion.

The energy of moving objects is called **kinetic energy**. The amount of kinetic energy (energy of motion) is found with the equation $KE = \frac{1}{2}mv^2$. This concept is closely related to Newton's first law of motion, that objects at rest tend to stay at rest and objects in motion tend to stay in motion. The greater the mass of an object and the greater its velocity, the harder it is to stop. This is called an object's **momentum**. A 5-ton truck going 80 miles an hour is a lot harder to stop than a kid on a skateboard going 3 miles an hour. It takes more energy to move a greater mass and it takes more energy to give it a greater velocity. The momentum of an object is found by $p = mv$. The letter p is used to stand for momentum. It would be confusing to use m, because m is already used for meters (m) and mass (m).

At times, an object has the potential to move but is not moving. An example is a book sitting on the edge of a bookshelf. Someone used energy to place the book on the shelf. All you must do is give the book a small push and it will fall to the floor. You do not have to pull or push it to the floor. Gravity will do that for you. It is already in the gravitational force field except it is prevented from falling by the shelf.

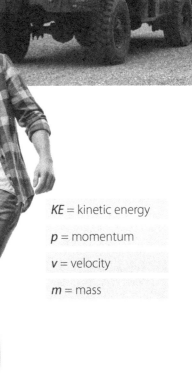

KE = kinetic energy

p = momentum

v = velocity

m = mass

The higher the shelf, the greater the movement when it does fall. The greater the mass of the book, the harder it will hit the floor. When it falls, it will accelerate at 9.8 $\frac{m}{s^2}$. The energy that moves the book to the floor when you give it a nudge is called its **potential energy** because the energy has the potential to be used when the obstacles are removed. The amount of energy used when the book falls to the floor is affected by g, its mass and its height. So, the equation for gravitational potential energy is $PE = mgh$. If the shelf is 1 meter above the floor and the mass of the book is 0.1 kg, the potential energy moves the book to the floor is $mgh = (0.1 \text{ kg})(9.8 \frac{m}{s^2})(1 \text{ m}) = 0.98 \text{ J}$.

PE = potential energy
m = mass
g = gravitation force
h = height
v = velocity
F = force
t = time of impact

Another example is a shelf that is 3 meters high and a book with a mass of 0.05 kg.

$$PE = (0.05 \text{ kg})(9.8 \tfrac{m}{s^2})(3 \text{ m}) = 1.47 \text{ Nm} = 1.47 \text{ J}.$$

For the brief instant when the book is pushed off the shelf, its potential energy is maximum, and its kinetic energy is zero because it has not begun to fall. When it falls, its velocity increases because it is being accelerated at 9.8 $\frac{m}{s^2}$. Its kinetic energy ($\frac{1}{2}mv^2$) increases and its height decreases, causing its potential energy (mgh) to decrease. Right before it hits the floor, its kinetic energy is at a maximum and its potential energy is virtually zero. When it hits the floor, it no longer has kinetic or potential energy. What happened? Did the energy just disappear? That is impossible. Energy is conserved, meaning that it cannot be created nor destroyed (except by God). The energy is transferred to deforming the floor or the book or both, or rapidly moving air molecules causing sound (thud).

The greater the momentum of the book, the harder it is to stop when it hits the floor. The momentum of the book becomes **impulse**. Momentum is mv and impulse is Ft (the force of the impact times the time of the impact). The units of momentum are mass × velocity which is kg × $\frac{m}{s}$. The units of impulse of force × time, which is N × s = $\frac{kg \; m}{s^2}$, which is kg × $\frac{m}{s}$ (because $\frac{ms}{s^2} = \frac{m}{s}$). If the kinetic energy right before impact is greater, the momentum will be greater, and the impulse upon impact will be greater. To put it another way, when an object hits an immovable object, the object ceases to move, and its momentum becomes impulse. This can be seen (and felt) when you catch a baseball bare handed. You should pull your hand back while catching the ball which would increase the time of impact of the ball with your hand. The momentum of the ball before impact determines the impulse. If the time of the impact can be increased (by pulling your hand back), the force will be reduced because the force times the time of impact equals the momentum of the ball before impact.

If a 0.25 kg ball hits your hand going 20 $\frac{m}{s}$ and the time of the impact is 0.1 s, what is the force of the impact on your hand?

$$mv = (0.25 \text{ kg})(20 \tfrac{m}{s}) = F \times (0.1 \text{ s})$$

$$F = \frac{(0.25 \text{ kg})(20 \tfrac{m}{s})}{0.1 \text{ s}} = 50 \text{ N}$$

What if, instead, you pull your hand back and the time of impact is increased to 0.5 s?

$$F = \frac{(0.25 \text{ kg})(20 \frac{\text{m}}{\text{s}})}{0.5 \text{ s}} = 10 \text{ N}$$

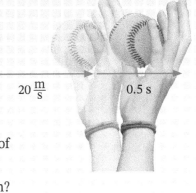

You increased the time of impact by 5 times which reduced the force of impact by 5. That could be very good for your hand. Remember that the next time you play catch. Just do not do it with a bowling ball. That would increase the mass of the ball which would increase its momentum and increase the impulse! Ouch!

What would be the potential energy of a 2 kg bowling ball on a shelf 1.5 m high?

$$PE = mgh = (2 \text{ kg})\left(9.8 \frac{\text{m}}{\text{s}^2}\right)(1.5 \text{ m}) = 29.4 \text{ Nm} = 29.4 \text{ J}$$

What is the maximum kinetic energy of the bowling ball when it hits the floor? It would also be 29.4 J because the potential energy becomes 0 when the height becomes 0, so all the energy is kinetic energy right before impact. What then is the velocity of the bowling ball upon impact?

$$KE = \tfrac{1}{2}mv^2 = 29.4 \text{ J and } v^2 = 2\frac{29.4 \text{ J}}{m}$$

$$v^2 = 2\frac{29.4 \text{ J}}{2 \text{ kg}} = 29.4 \frac{\text{m}^2}{\text{s}^2} \text{ and taking the square root of this result gives}$$

$$v = 5.4 \frac{\text{m}}{\text{s}}$$

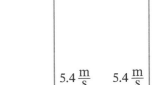

The momentum of the bowling ball just before impact is

$$mv = (2 \text{ kg})\left(5.4 \frac{\text{m}}{\text{s}}\right) = 10.8 \text{ J}$$

If the floor is very rigid and the time of impact of the bowling ball and the floor is 0.1 s, the force of impact is

$$F \times (0.1 \text{ s}) = 10.8 \text{ J}$$

$$F = \frac{10.8 \text{ J}}{0.1 \text{ s}} = 108 \text{ N}$$

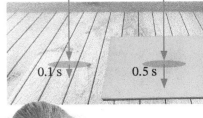

What if the floor was cushiony and increased the time of impact of the bowling ball to 0.5 s?

$$F = \frac{10.8 \text{ J}}{0.5 \text{ s}} = 21.6 \text{ N}$$

That is sure a lot easier on the floor.

If you are having trouble following the algebra, focus on how the problems are solved and the understanding will come later. In the example above, where $F \times (0.1 \text{ s}) = 10.8 \text{ J}$, both sides of the equation were divided by 0.1 s. The left side of the equation became $F \times \frac{0.1 \text{ s}}{0.1 \text{ s}}$, which is F. The right side of the equation became $\frac{10.8 \text{ J}}{0.1 \text{ s}}$, which is 108 N. You will find more examples in the worksheets.

How does a person break a stack of boards or blocks with a bare hand without the hand shattering? You may have seen a demonstration of this martial arts feat. The person must bring the hand down with a high velocity producing a large momentum. The momentum must be in the velocity because the mass of a human hand is quite small. Then the hand must hit the boards or blocks with very small time of impact. This means that the hand must be withdrawn very quickly. This takes considerable training and discipline. Do not just go out and try it yourself.

In **elastic collisions**, the colliding objects are not permanently deformed by the collision. They snap back to their original shape, hence the word *elastic*. (The opposite of this is called inelastic collision.) Their kinetic energy and momenta (plural of momentum) are conserved. No energy is used to deform the colliding objects. Too bad that that does not happen when two cars collide with each other. If a fast-moving object collided in an elastic collision with a slower-moving object, the faster moving object would go more slowly, and the slower object would go faster. If you added the momenta of the objects before the collision and after the collision, they would be the same. This is the **law of conservation of momentum**. A law in science is based upon many observations like the Law of Gravity.

Air escaping a balloon and the balloon flying around a room was used in chapter 3 as an example of Newton's Third Law of Motion. It can, perhaps, better be looked at as an example of the conservation of momentum. When the air was in the balloon, the mass and velocity of the balloon and the air made up the mass of the same object. As the air escaped the balloon, the escaping air had a separate mass and velocity in the opposite direction of the balloon. The mass of the escaping air times its velocity is the momentum of the escaping air. The momentum of the escaping air and the momentum of the flying balloon add up to equal the momentum of the original balloon with air. The escaping air has a negative momentum because it is going in the opposite direction as the balloon. The momentum of the flying balloon increased because it has a greater velocity.

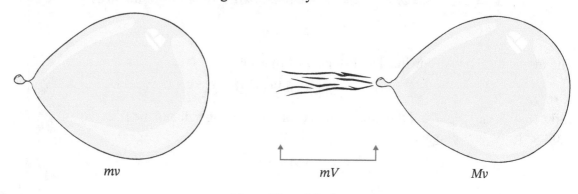

$Mv - mV$ = original mv

mV is negative because it is in the opposite direction

A similar situation exists for a jet with the thrust going out in the direction opposite to the motion of the jet.

$Mv - mV$ = original mv

Mv = momentum of jet

$-mV$ = momentum of exhaust

$Mv - mV$ = original mv

This is also the case of a spacecraft in outer space where the exhaust has nothing to push off from. Here the momentum of the spacecraft before emitting the exhaust equals the sum of the momenta of the spacecraft and the exhaust separately.

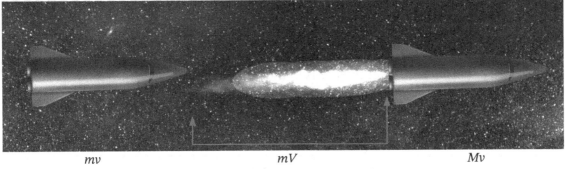

$mv \qquad mV \qquad Mv$

$Mv - mV = $ original mv

An example of an elastic collision would be a ball that hits a hard surface. The ball is crushed on the side that hits a hard surface and returns to its original shape after it bounces back.

What is the force of impact when a 3,000-pound car going 60 miles an hour hits a block wall and the time of impact is 0.1 s? First you must find the momentum of the car.

Pounds are units of force in the English system. To find the momentum of the car we need to know the mass of the car. Pounds are units of force and kilograms are units of mass. We know that when the acceleration of gravity is $9.8 \, \frac{m}{s^2}$, a mass of 0.4536 kg has a weight of 1 pound. A 3,000-pound car has a mass of

$$3{,}000 \times 0.4536 = 1{,}360.8 \text{ kg}$$

To convert 60 miles per hour into metric units, 1,609 meters in a mile:

$$60 \text{ miles} \times 1{,}609 \, \frac{\text{meters}}{\text{mile}} = 96{,}540 \text{ meters}$$

There are 3,600 seconds in an hour because there are 60 minutes in an hour and 60 seconds in a minute.

The velocity of the car is

$$\frac{96{,}540 \text{ m}}{3{,}600 \text{ s}} = 26.8 \, \frac{\text{m}}{\text{s}}$$

$$mv = Ft \text{ and } \frac{mv}{t} = F$$

$$\frac{(1{,}360.8 \text{ kg})(26.8 \, \frac{\text{m}}{\text{s}})}{0.1 \text{ s}} = 364{,}694.4 \text{ N}$$

To convert Newtons into pounds there are 4.448 N in a pound.

$$\frac{364{,}694.4 \text{ N}}{4.448 \, \frac{\text{N}}{\text{lb}}} = 81{,}990.6 \text{ lb}$$

That may total the passenger as well as the car.

If the velocity of the car is 30 miles per hour instead of 60 miles per hour, the momentum is cut in half and so is the force of impact. If the wall gives way, increasing the time of impact, the force of impact is reduced.

This is why it is always said that speed kills. Always remember this when you get behind the wheel of a car.

When you go up a flight of stairs, you are doing work (you are moving yourself) and it takes energy. Your muscles exert force to overcome the force of gravity. When you get to the top of the stairs, your potential energy is *mgh*. This is the energy it took to get you to the top of the stairs. Would you feel any different if you slowly went up the stairs or ran up the stairs? Of course, you would. You would use the same amount of energy (*mgh*) in each case. So, what is different? The difference is in the time it takes you to go up the stairs. If you took the amount of work performed (or energy used, *mgh*) and divide it by the time it took to use the energy, you would get **power**. Each of these concepts were originally developed by looking at situations in everyday life. It is a matter of observing and trying to make sense of the circumstances of life. When you go up the stairs, the power you use is the energy (or work performed) divided by time.

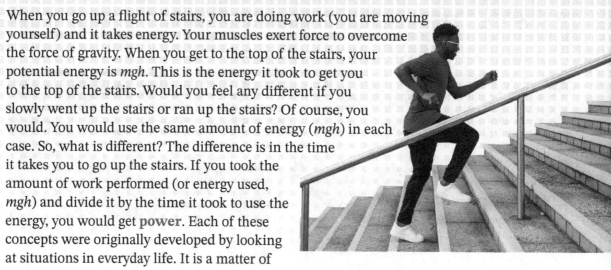

$$P \text{ (power)} = \frac{mgh}{t}$$

If you took more time to go up the stairs, you would divide *mgh* by a larger number and your power would be less. If you ran up the stairs, your time would be smaller, and the power would be greater. The work performed (energy used) is in units of joules and the time is in seconds, so the power would be $\frac{J}{s}$ which is in units of **Watts (W)**. Whether you go up the stairs fast or slow, you use the same amount of energy (*mgh*). But if you go up fast, you feel more tired afterward.

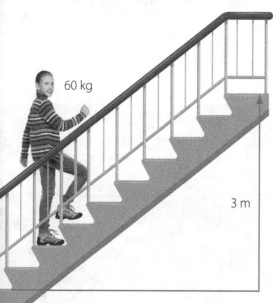

How much power would you use if you went up the stairs to a height of 3 meters in 2 minutes if your mass is 60 kg?

Knowing that you want to find $\frac{J}{s}$, you need to convert the 2 minutes into seconds. There are 60 seconds in a minute, so

$$2 \text{ minutes} = 2 \text{ minutes} \times \frac{60 \text{ seconds}}{\text{minute}} = 120 \text{ seconds}$$

When you do calculations like this, ask yourself if your answer makes sense. What if you had done the calculation as $\frac{2 \text{ minutes}}{\frac{60 \text{ seconds}}{\text{minute}}} = 0.03$ second. You know that there should be more seconds than minutes, so 0.03 seconds must be wrong.

The power used in going up the stairs is

$$P = \frac{mgh}{t} = \frac{(60 \text{ kg})\left(9.8\frac{m}{s^2}\right)(3 \text{ m})}{120 \text{ s}} = 14.7 \frac{J}{s} = 14.7 \text{ W}$$

What if you raced up the stairs in $\frac{1}{2}$ minute? Half of a minute is 30 seconds, so

$$P = \frac{(60 \text{ kg})\left(9.8\frac{m}{s^2}\right)(3 \text{ m})}{30 \text{ s}} = 58.8 \text{ W}$$

58.8 W is more power than 14.7 W. You would use more power going up the stairs faster and you would feel it.

When James Watt invented the steam engine, people wanted to know how long it would take the steam engine to do the same amount of work as a horse. He calculated that a healthy horse could do 550 foot pounds (English units of work, force times distance) of work in a second. From this he defined the unit of power known as the **horsepower** (hp) where 1 hp equals 550 foot pounds per second. In metric units, 746 W are equal to 1 hp and a kW (kilowatt or 1,000 W) equals $1\frac{1}{3}$ hp.

When you went up the stairs in 30 seconds, how many hp did you use? You used 58.8 W and 1 hp is 746 W.

$$\frac{58.8 \text{ W}}{746 \frac{\text{W}}{\text{hp}}} = 0.08 \text{ hp}$$

You used 8 hundredths of the power that a healthy horse would have used. Of course, this value is an average for many sizes and types of horses. Would it then be an advantage to ride a horse up the stairs? Maybe that is not such a bright idea. Horses can be kind of messy on stairs.

If a car is rated at 280 hp, how much power is that in kW?

$$(280 \text{ hp})\left(746 \frac{\text{W}}{\text{hp}}\right) = 208{,}880 \text{ W} = 208.9 \text{ kW}$$

Horsepower comparisons

Motorcycle: 80 to 100 Hp

Small car: 180 to 200 Hp

Large SUV: 300 Hp

Semi truck: 400 to 600 Hp

Notice that you move the decimal to the right 3 places because there is 1,000 W in a kW.

If your mass is 70 kg and you ride up a hill 4 m high in a 40 kg wagon in 5 minutes, how much power did you use?

$$5 \text{ minutes} \times 60 \frac{\text{seconds}}{\text{minute}} = 300 \text{ seconds}$$

The total mass moved up the hill is 70 kg + 40 kg = 110 kg.

$$P = \frac{mgh}{t} = \frac{(110 \text{ kg})\left(9.8 \frac{\text{m}}{\text{s}^2}\right)(4 \text{ m})}{300 \text{ s}} = 14.4 \text{ W}$$

Energy is a good example of a very fundamental part of the created universe that we cannot see but can measure by its effects upon matter. Energy can be stored as potential energy and used to move objects as kinetic energy. To say that something has to be seen to be real denies the very nature of physics.

LABORATORY 6

Kinetic Energy, Potential Energy, Momentum, Collision, and Power

REQUIRED MATERIALS

- Small object like a toy car with wheels
- Meter stick
- Piece of string
- A weight to place on the car (this can be anything that will stay on the car)
- Spring scale that measures Newtons
- A board to use as a ramp
- Protractor
- Book on a shelf or ledge
- Tennis ball or other bouncing ball
- **Note:** An assistant is needed to help with this lab.

Introduction

When you move an object, you are doing work on the object. It is calculated as the force applied to the object as measured in Newtons times the distance that the object is moved as measured in meters. The amount of energy required to move an object is the amount of work in moving the object. This is a mechanical concept of work and energy. There are also applications in dealing with electricity and light.

Sometimes when you move an object, the energy used to move it is stored as potential energy such as when you place a book upon a shelf from which it could fall. When the book can fall from the shelf, its potential energy becomes kinetic energy of motion until it hits the floor.

Power is a measure of how much energy is used over a given period. Energy is measured in Newton meters (Nm) which are joules (J). Power is a measure of $\frac{J}{s}$. When greater power is used, more energy is used over a shorter period — such as when you run upstairs instead of walking.

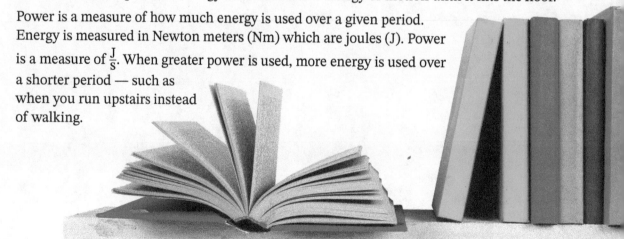

Purpose

The purpose of this lab exercise is to measure work and energy; measure potential and kinetic energy of a falling object; demonstrate elastic and inelastic collisions; and observe the horsepower of common objects in daily use and convert their horsepower to $\frac{J}{s}$.

Procedure

 observe

1. Find a small object with wheels (such as a toy car) that you can pull with a string. Attach a piece of string to the object and pull it a short distance. The purpose of using an object with wheels is to minimize the impact of friction. Attach a spring scale (that measures Newtons) to the string. Pull on the scale so that the object moves 2 meters (measured with a meter stick) in a straight line. Observe and record the number of Newtons on the scale required to pull the object. Repeat the procedure with additional weight added to the object.

 question

2. How much work was done in moving the object each time? The work performed is the Newtons of force used times the distance the object was pulled (2m).

research

3. Take a board and make a ramp (called an inclined plane) so that you can pull the object uphill. Place the ramp at about a 30° angle as measured with a protractor. Do this without and with the added weight. Calculate how much work you performed in each case.

4. Determine the mass (with a digital scale) of a book. Place the book up on a shelf or ledge above the floor. Measure the height of the shelf in meters. Calculate the gravitational potential energy of the book ($PE = mgh$) on the shelf. Push the book off the shelf so that it falls to the floor. What was the potential energy and kinetic energy of the book when it was halfway to the floor? When the book is halfway to the floor, its height is half as much, so its potential energy is half as much. The potential energy became kinetic energy, the kinetic energy is equal to the potential energy. What is the potential energy right before impact? The height should be essentially zero, so the potential energy should be zero. The kinetic energy should be maximum and equal to the original potential energy of the book before it fell.

5. Drop a book and a heavier object on the floor. Describe the sound of each as it hits the floor. Describe the differences in their impacts as they hit the floor. Which has the greater impulse? How is this related to their masses?

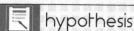 hypothesis

6. This procedure will take 2 people. Have someone drop a tennis ball and you are to observe it hitting the floor or the ground. You must get down so that you can see what happens to the tennis ball as it hits the floor without it hitting you. You could do this with a larger ball that will bounce as well. When the ball hit the floor, was it an elastic or inelastic collision? Be sure to review this concept in this chapter. Explain why it was an elastic or inelastic collision. Explain this in detail using complete sentences.

7. Take the ball and drop it so that it bounces from the floor and allow it to keep bouncing. Let it bounce until it stops bouncing.

experiment

8. Will it keep bouncing indefinitely? Why or why not? Answer the question again — is the collision of the ball with the floor an elastic or inelastic collision? Explain your answer in detail. If you dropped a lump of clay on the floor, would it be an elastic or inelastic collision? You can picture this in your head. You can try it if you have a lump of clay handy, but you do not need to go out and buy one.

 analyze

9. Look at a lawn mower and write down its horsepower. If you do not have one, you can observe one at a local hardware store. Convert its horsepower to watts.

10. Look up the horsepower of a large car and a smaller compact car. Convert each to watts.

conclusion

11. What would a large heavy car with lower horsepower be like? What about a large heavy car with a great deal of horsepower? What would a small, lighter car with lower horsepower be like? What about a small, lighter car with a great deal of horsepower? Which car would you like to have? Why?

High hp

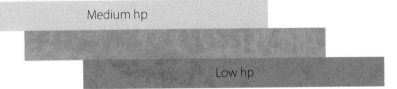

Medium hp

Low hp

Laboratory 6 | 57

CHAPTER SEVEN

WORK, MACHINES, AND TORQUE

OBJECTIVES

At the conclusion of this lesson students should have an understanding of

- Work
- Machines and mechanical advantage
- Inclined plane
- Lever and fulcrum
- Gears and pulleys
- Torque lever arms

58 | Physics

How much work would have to be done to lift a 100-pound safe up 2 feet to get it into the back of a truck? Work is force times distance, so the answer is . . .

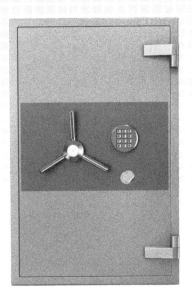

$$100 \text{ pounds} \times 2 \text{ feet} = 200 \text{ foot pounds}$$

Foot pounds are the units of work in the English system.

It would be a difficult task to lift the safe straight up into the truck unless you are in very good shape. But even then, it is very easy to create lifetime back problems. This task can be done much easier with a machine. It does not have to be something with a motor and a hoist but can be as simple as a board. Anything can be a **machine** if it can reduce the amount of force required to do the work and/or change the direction that you exert the force.

A ramp in the back of a truck is a machine called an inclined plane (a plane is a flat surface like a board). It is a simple board or similar object upon which you can push something. If you lift the safe straight up, you must lift 100 pounds. If the ramp in the back of the truck is 20 feet long, you can push the safe (hopefully it has wheels) 20 feet up a gradual incline with a force of only 10 pounds. The machine (ramp) gives a **mechanical advantage**. Divide the distance of the ramp (20 feet) by the straight up distance (2 feet) and you get a mechanical advantage of 10. This means that you only must push the safe up the ramp with a force of $\frac{100}{10}$ or 10 pounds using less force over a greater distance. That is much easier and better on your back.

It is important to understand that you are not getting something for nothing. Energy and work do not work that way. You still must put as much potential energy into the safe either way because it still ends up 2 feet above the ground. With the ramp, you exert 10 pounds over 20 feet instead of 100 pounds over 2 feet. The potential energy is mgh and the mass of the safe and final height are the same both ways.

$$\text{Work} = F \times d$$
$$= 100 \text{ pounds} \times 2 \text{ feet} = 10 \text{ pounds} \times 20 \text{ feet} = 200 \text{ foot pounds}$$

A teeter-totter on the playground is a machine called a **lever**.

The part in the middle that supports the teeter-totter is the **fulcrum**. To apply it as a machine, imagine that you came upon a 300-pound rock blocking the road that you had to move to get by. If you could just lift it up from the ground, you could roll it out of the way. So, you quickly send away for a muscle building kit. Not very practical. You would probably look funny with those kinds of muscles anyway. Here is how it could work.

Find a smaller rock to use as the fulcrum. Then find a strong board at least 6 feet long. They are always found alongside the roadway (especially in the movies). Wedge part of the board under part of the big rock with about 1 foot of the board on that side of the smaller rock and 5 feet of the board on the other side of the smaller rock. Now when you push down $2\frac{1}{2}$ feet on the longer side of the board with 60 pounds of force (sit on it), you are doing enough work to lift the 300 pound rock $\frac{1}{2}$ foot so that you could roll it out of the way and continue on your journey down the road.

Diagram 7.1

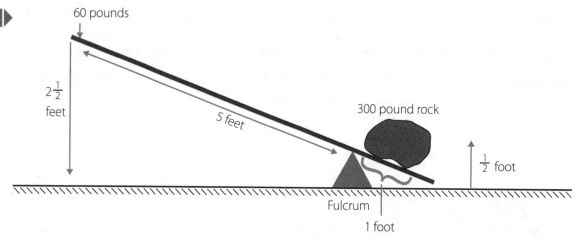

Another way to state it is that the work you do on the machine, the machine will do on something else. Or as it is sometimes expressed . . .

work in = work out

Being as work = force × distance, you can say that . . .

$$(2.5 \text{ feet})(60 \text{ pounds}) = (0.5 \text{ feet})(300 \text{ pounds})$$

$$150 \text{ foot pounds} = 150 \text{ foot pounds}$$

If you exert 60 pounds to move the machine $2\frac{1}{2}$ feet, the machine will exert 300 pounds to move the rock $\frac{1}{2}$ foot. Sounds fair enough. Here the mechanical advantage is. . .

$$\frac{300 \text{ pounds}}{60 \text{ pounds}} = 5 \text{ or } \frac{2\frac{1}{2} \text{ feet}}{\frac{1}{2} \text{ foot}} = 5$$

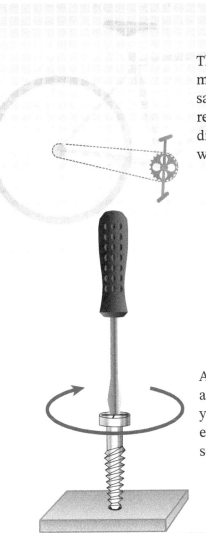

The gears on a bicycle are another machine. The larger gear turns in the same amount of time as the smaller gear but requires less force because you go a greater distance farther around. That is especially helpful when you are going up a hill.

A screw is another machine. As you turn a screw, you exert a force over a greater distance (going around the coils) than if you tried to push the screw directly into wood or metal. You are exerting less force over a greater distance and can accomplish something that might otherwise be impossible.

Diagram 7.2

Diagram 7.3

A pulley is also a machine. A pulley can change the direction of the force you have to exert without changing the force. Its mechanical advantage is 1. It would be easier to pull on a rope to lift a 50 pound object than to bend over and pick it up. It could save your back and knees. (*Diagram 7.2*)

If you use a setup of 5 (called a block and tackle), you would pull on the cord 5 times as far as the weight rises. Here you have a mechanical advantage of 5. The cord you are pulling on is wrapped around the 5 pulleys, so you are pulling it 5 × farther than the object you are lifting. Notice that the cord you are pulling is not directly connected to the weight. The hook on the mechanism connecting the pulleys connects to a cord attached to the weight. This arrangement is commonly used to lift an engine out of a car or truck in a mechanic's garage. (*Diagram 7.3*)

Chapter 7 | 61

A pair of scissors and a pair of pliers are machines. The blades of the scissors and the claws of the pliers move a shorter distance and exert more force than the longer handles where you apply the force.

In these machines, the fulcrum is a point around which the handles rotate as in the teeter-totter. This tendency to rotate is called the **torque**. A greater tendency to rotate means that it rotates easier. If the torque is greater, it is easier to rotate. The torque (τ, the Greek lower-case letter tau) is the force applied perpendicular to the lever times the distance from where the force is applied to the fulcrum. In the example used earlier... (*Diagram 7.4*)

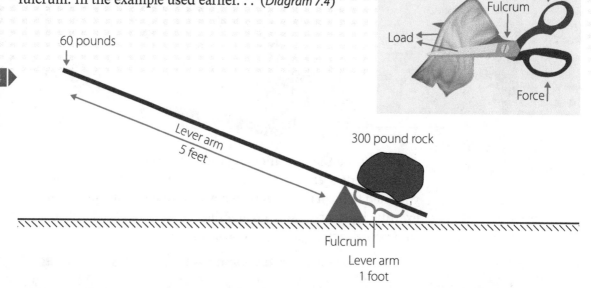

On the left side of the fulcrum, $\tau = $ (60 pounds) $\times$ (5 feet) $=$ 300 foot pounds, and on the right side $\tau = $ (300 pounds) $\times$ (1 foot) $=$ 300 foot pounds. If you increased the length on the left side to 10 feet, it would be easier to push down, and it would be easier for the board to rotate around the fulcrum because $\tau = $ (60 pounds) $\times$ (10 feet) $=$ 600 foot pounds.

The result of torque is motion in a circle. In this example, the rock did not go straight up and the force on the left side was not straight down. The board rotated around the fulcrum (smaller rock) in a circle.

In equation form, torque is expressed as . . .

$$\tau = r \times F$$

where τ is the torque, r is the radius of rotation which is the lever arm, and F is the perpendicular force.

If you try to turn a bolt with a wrench with a handle 6 inches long, you can make it easier by putting a piece of pipe a foot long over the handle of the wrench. This increases the length of the lever arm — the distance from where the force is applied and the fulcrum.

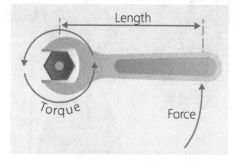

An auto mechanic uses a torque wrench to tighten the bolts that hold 2 pieces of an engine together. There is a gasket between those 2 pieces which is material to seal the gap between the pieces of metal. If the bolt is tightened too tight the gasket will be destroyed, with oil leaking out between the pieces of metal. The torque wrench measures the torque applied so that enough but not too much torque is applied.

Newton's first law of motion stated that an object at rest tends to stay at rest unless unbalanced forces act upon it. In this case, the balance of equal torques on both sides of the board cause the board to remain horizontal. If there is a bit more torque on one side or the other, the board will rotate in the direction of the greater torque. Notice that this is not necessarily in the direction of greater force but greater torque. The torque on one side can be increased by increasing the length of the lever arm on that side without changing the force. in the example of the 300-pound rock, a little greater torque on the left side will cause the right side to go up. When the torques on both sides are the same, it is in equilibrium and it will not rotate.

A tight rope walker in a circus uses a long pole for balance. This provides a long lever arm on both sides which means that it does not take much force to balance the torque on each side.

A hammer is a very useful machine. Can you imagine trying to pull a nail out of a board by grabbing it with your fingers and pulling on it? The lever arm of a hammer is the length of the handle to where the head of the hammer touches the board. The length of the handle to the fulcrum (point on the metal part of the hammer that touches the board about which the hammer rotates) times the force you apply to the handle equals the distance from the fulcrum to the nail times the force applied to the nail (*Diagram 7.5*).

$$F \times r + F \times r$$
(lever arm from fulcrum to the nail)

If the nail is still too hard to pull out of the board, slip a piece of metal pipe over the handle of the hammer. This will increase the lever arm on the side of the handle, increasing the torque and increasing the tendency of the hammer to rotate and pull out the nail.

I had to get a heavy mirror in a wooden frame upon a dresser by myself. The mirror and its frame were too heavy to lift so I leaned the frame against the dresser and pushed it up onto the dresser. The wooden frame containing the mirror became its own inclined plane. I pushed it up at an angle instead of straight up. It was much easier. The total distance that I pushed the frame with the mirror divided by the height of the dresser was the mechanical advantage. These concepts can save you a lot of physical misery throughout life. Knowing how to use knowledge is wisdom (*Diagram 7.6*).

Machines and leverage are fantastic gifts from God. Work equals force times distance and less force is required when the distance something is moved is increased. Remember this the next time you have to move a heavy object. It can save your back, legs, and knees a lot of future pain!

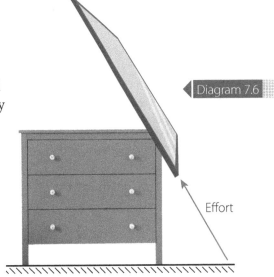

LABORATORY 7

Machines And Mechanical Advantage

REQUIRED MATERIALS

- Tape measure
- Trip to a hardware store to find the mechanical advantages of an appliance dolly, a wheelbarrow, and another machine.
- Meter stick
- Spring scale that measures Newtons
- 2 single pulleys
- 100-gram weight

Introduction

A machine is a device that either changes the direction of a force or reduces a force. For example, 1 pulley will redirect a force exerted on an object. If you wrap a cord around a pulley connected to an object you want to lift, you can pull down on the cord instead of having to pull up on it. That could be a lot easier and allow you to use your body weight to pull it.

If you use a 5 pulley block and tackle, you pull down the cord 5 times as far as an object is lifted. You have a mechanical advantage of 5 so you only must exert $\frac{1}{5}$ the force of the weight of the object.

With a machine, you can push or pull an object with a lot force. If you are using a lever, you push down on one end and the other end pushes up on the object you want to lift. If you push on an end with a lever arm 5 times longer than the lever arm on the end you want to lift, the object lifted goes up $\frac{1}{5}$ as high but with 5 times the force.

Purpose

This lab exercise increases the student's understanding of common machines and provides practical experience in applying mechanical advantages to everyday life.

observe

Take a tape measure, note pad, and pencil and go to a local hardware store for steps 1–3.

1. Find an appliance dolly (you can do this at home if you have an appliance dolly) and measure the length from the handles to the wheels (fulcrum) and the length from the wheels to the ledge that the appliance sits on. These can be measured in inches or cm (centimeters). Calculate the mechanical advantage of the dolly.

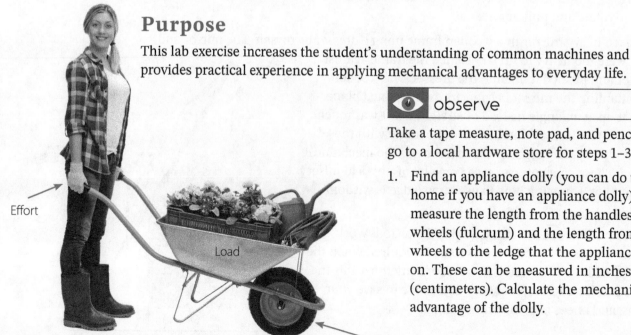

64 | Physics

question

2. How much force would it take to lift a 50-pound refrigerator with the dolly?

observe

3. Find a wheelbarrow (you can do this at home if you have a wheelbarrow) and measure the length from the handles to the wheels (fulcrum) and the length from the wheels back up to the middle of where the load would be. These can be measured in inches or cm (centimeters). Notice that this is a machine where both lever arms are on the same side of the fulcrum. Calculate the mechanical advantage of the wheelbarrow.

question

4. How much force would it take to lift 200 pounds of cement mix with the wheelbarrow?

observe

5. Find another machine in the hardware store, describe it, and calculate its mechanical advantage.

question

6. Use complete sentences to describe how you calculated the mechanical advantage.

research

7. Use a spring scale to weight the 100-gram weight in Newtons. The scale can be attached to a hook on the weight.

8. Set up the 2 pulleys as a block and tackle as shown in this diagram. Hang 1 pulley from an eye hook attached to a secure surface. Tie a piece of string to the lower hook on the pulley. Attach the 100-gram weight to a hook on the other pulley. Wrap the string from the first pulley around the second pulley and back around the first pulley as in the diagram. Attach the spring scale to the end of the string and pull it to lift the weight.

hypothesis

9. Record how many Newtons of force you are using to lift the weight with the pulleys. What is the mechanical advantage of using the pulleys?

experiment

10. Divide the force needed to lift the weight by itself by the force used with the pulleys.

analyze

11. Was the mechanical advantage what you expected?

conclusion

12. Why or why not? Use complete sentences.

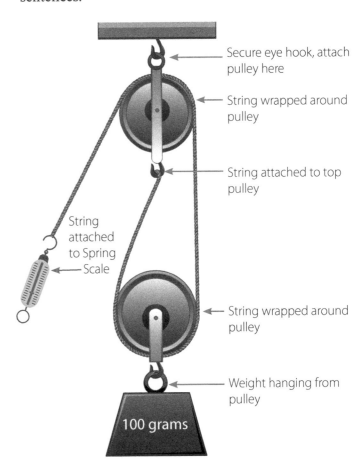

Laboratory 7 | 65

CHAPTER EIGHT

ROTATIONAL MOTION

OBJECTIVES

At the conclusion of this lesson the student should have an understanding of

- Revolution and rotation
- Translational and rotational velocity
- Application of torque
- Rotational inertia and angular momentum
- Conservation of angular momentum
- Centripetal and centrifugal force
- Tangential motion
- Angular velocity and angular acceleration

Motion in a circle is very common. Whether you are rotating a wrench to remove a bolt or steering a car around a curve, you have rotational motion. Any time you change direction, you have rotational motion.

Circular motion can be either spinning on an axis (rotating) (*Diagram 8.1*) or spinning around an external axis (**revolution**) (*Diagram 8.2*).

Earth rotates on its axis every 24 hours and revolves around the sun (external axis) every year. God has created each planet uniquely. Venus rotates around its axis in the opposite direction as earth (called **retrograde motion**) every 243 earth days and revolves around the sun every 225 earth days. Mercury has a more unusual pattern in that it rotates every 59 earth days and revolves around the sun every 88 earth days. The result is that 1 day on Mercury as measured from noon to noon is 2 years on Mercury. Creative design was placed upon every planet. If this were not the case, science could be very boring. It is extremely difficult for us as humans to conceive of a God that is infinite and spoke everything into being. When you see through the study of physics and the other sciences that the universe is like a large machine that seems to have an infinite number of pieces all continually working together in great precision, it is hard not to see an all-knowing God behind it all. I suspect that many atheists have not looked at the detail in this universe we live in. Notice as you go through these chapters that many of the concepts and applications seem to pop up several times in different ways. They are different ways of looking at the same thing.

Diagram 8.1
Rotation around internal center

Diagram 8.2
Revolution around external center

> **The heavens declare the glory of God; and the firmament shows His handiwork (Psalm 19:1).**
>
> **I will meditate on the glorious splendor of Your majesty, and on Your wondrous works (Psalm 145:5).**

Astronomy is part of physics and includes many interesting applications of the principles you are studying. There are times in our lives when we are drawn away from God and His Word. The heavens can remind us of His wisdom shown in His creation and Word.

The moon revolves around earth and rotates at the same rate, resulting in the same side of the moon always facing us. God has gone out of His way to show His design in the universe.

The circumference divided by the diameter of a circle is a constant number called **pi** (π). It is 3.14. . . . The numbers keep going to the right of the decimal. For our use, we will call it 3.14. This is $\frac{C}{d} = \pi$. It is much easier to measure the diameter of a circle with a ruler than to try to measure the circumference around a circle. Like many other things in physics, it is easier to calculate the circumference than to measure it. With a little algebra, the equation becomes $C = \pi d$.

In discussing rotational motion, it is important to distinguish between **translational velocity** around a circle measured in meters per second and **rotational velocity** as measured in RPM (revolutions per minute). A spinning DVD has the same rotational velocity throughout the disc. Since it is a solid disc, the outer rim of the disc spins around the same number of times in a minute as the inner part

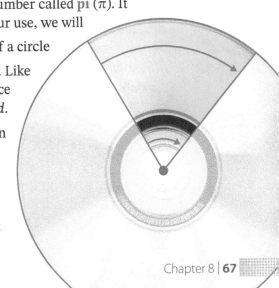

of the disc. However, the translational velocity of the outer rim on the disc is greater than the inner part of the disc because the distance is greater around the outer rim.

It may not seem obvious at first, but a swinging pendulum also has rotational motion — it is just not swinging all the way around the circle. When you walk, your legs and arms also act like a pendulum going back and forth through a partial circle. The length of your legs is the length of a rotating radius. If you have longer legs, the radius of rotation is longer and the velocity of the back and forth movement of your legs tend to be slower. Because your feet are farther from your body, their translational velocity in $\frac{meters}{second}$ is greater than that of your hips. Long-legged critters tend to have a slower gait to keep their feet from going so fast. Therefore, long-legged animals, like a giraffe, have a slower gait than a short-legged animal like a dachshund.

Newton's First Law of Motion is that a moving object will keep moving unless there is an opposing force, and a stationary object will not move unless there is a force to make it move. This is also true for rotating objects which will keep rotating unless there is an opposing force. As well, objects that are not rotating will not rotate unless there is a force causing the rotation. The force that causes or reduced the rotation is part of **torque** (τ) that was introduced in chapter 7. Recall the example of moving the 300-pound rock by using a lever. A board was placed under the rock. The middle of the board was on a smaller rock called a fulcrum. The side of the board on the other side of the fulcrum was longer. Force times distance was the work performed on each side of the fulcrum.

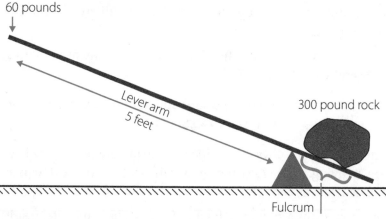

The left side of the board is 5 feet long, so it took 60 pounds to balance the 300 pounds on the right side. The board and fulcrum are a machine called a lever. The length of the board on each side of the fulcrum is the arm of the lever — hence, the name "lever" arm. The portion of the board on the right side had to rotate counterclockwise and the portion of the board on the right side had to rotate clockwise. Torque is a force perpendicular to the lever arm times the length of the lever arm, which is the work it takes to rotate the lever.

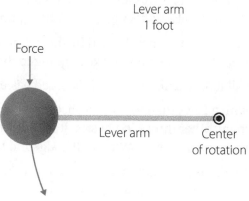

Think of a ball on the end of a rod that is connected to a support in the middle of a circle. When a force is applied to the ball at a 90° angle (perpendicular) to the rod, it will revolve around the circle. The force multiplied by the length of the rod is the torque that causes the ball to revolve about the circle.

When the ball is revolving around the circle, it will take another force in the opposite direction perpendicular to the rod (lever arm) to slow it down. The larger the radius of the circle, the faster the ball will revolve because it has farther to go around a larger circle in the same amount of time.

The momentum of an object going in a straight line is mv because the mass and velocity contribute to its inertia. For the ball going in a circle, the radius also has an effect because a larger radius means a larger lever arm, which means greater torque. This is called **rotational inertia.** This results in a momentum (called **angular momentum**) that keeps the ball going in a circle. If the ball goes a quarter of the way around the circle, it travels through an angle of 90°. As the ball goes around the circle, its distance can be measured in angles, so it is called angular. Since the mass, velocity, and radius all affect the angular momentum, the angular momentum is mvr. The v (velocity) in this expression is translational velocity measured in meters per second, which is perpendicular to the radius. As the ball revolves, the velocity keeps changing direction around the circle.

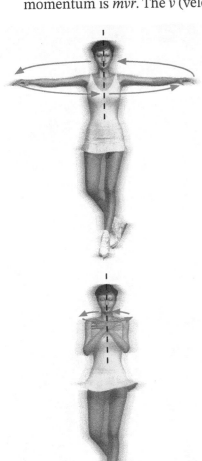

Like the law of conservation of momentum, which was presented in chapter 6, angular momentum is also conserved if there is no additional torque applied to the rotating or revolving object. Consider the example of a spinning ice skater. If the skater extends her arms, her radius (r) of rotation increases. The mass of the skater hopefully does not change, so v must decrease to keep mvr constant, so she spins more slowly. When she pulls her arms close to her body, her radius decreases, v increases, and she spins faster. Therefore, her velocity of rotation appears to increase and decrease without applying any other effort.

When an object is revolving around another object, like the earth around the sun, there must be a force pulling the revolving object toward the center of the circle or it will fly off. Gravity between the sun and earth keeps earth orbiting around the sun. This is called a **centripetal** (center seeking) **force**. In an earlier lab exercise, you put water in a bucket and swung it over your head. Your arm and hand holding the handle of the bucket provided the centripetal force. When you swung it over your head, another force caused the water to stay up against the bottom of the bucket instead of giving you a shower. This force came from Newton's third law of motion, which states that for every force there is an equal and opposite force. This force opposite the centripetal force, called the **centrifugal** (meaning center fleeing) **force**, kept the water against the bottom of the bucket. Some call it a fictitious force because it is a reaction rather than being caused by something like gravity. Either way, it acts on the revolving object. If the handle of the bucket broke, there would no longer be centripetal force and centrifugal force. The translational velocity of the bucket would cause it to fly off in a straight line called the tangent to the circle. This is what happens when you throw a baseball. While you

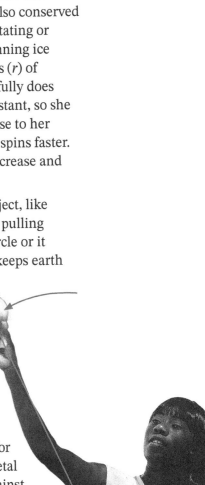

hold onto the baseball, your arm provides the centripetal force. When you release the ball, it flies off in a straight line. Its direction is in the direction of the torque which is at a right angle to the radius (lever arm). The motion of the bucket or baseball after it is released is called its **tangential motion**.

Some have planned a future space colony that is a large circular rotating tube. Astronauts would be held to a flat surface along the inside of the outer rim by centrifugal force. The centripetal force is supplied by the tubes connecting the outer tube to the center of the space colony. This is an example of simulated gravity. The astronauts can function and move about as if they were experiencing actual gravity. Sometimes there are some small, odd symptoms experienced because an astronaut's head is closer to the center of the space station than the astronaut's feet. This causes the torque to be a little less than on the feet.

Centrifugal force stimulates gravity on astronaut.

Structures and large molecules (proteins, lipids, starches, etc.) from within cells are separated from each other with an ultracentrifuge that produces enormous centrifugal forces to pull heavier structures and molecules away from lighter ones that remain suspended in fluid in a revolving tube. The very large rates of revolution produce centrifugal forces thousands of times the force of gravity experienced at sea level. A centrifugal force 1,000 times the force of gravity is written as 1,000 g.

Material pushed to bottom of the tube by centrifugal force.

Sometimes it is more advantageous to use angular velocity, such as revolutions per minute, instead of translational velocity. If a tiny insect were on the outer rim of a DVD, it would go around as many times as an insect closer to the middle of the DVD. A satellite orbiting the earth may have an orbit farther from earth than another satellite. But for communication purposes it may be necessary that they both orbit around earth together. The standard unit of distance around a circle is a **radian**. It is the length of arc an object travels around a circle divided by the radius of the circle. Remember that the distance around a circle (circumference) divided by the diameter is π. The diameter is twice the radius, so $\frac{C}{2r} = \pi$, which gives $C = 2\pi r$. Therefore, there are 2π radians around a circle. If an object

Ultracentrifuge isolates heavier cellular materials from lighter materials.

has a rotational velocity of $2\pi \frac{radians}{second}$, it goes around the circle once every second. If an object goes $\pi \frac{radians}{second}$, it goes halfway around the circle every second. To convert translational velocity (v) in meters per second into rotational velocity (v) radians per second, use the equation $v = \frac{v}{r}$. An orbiting object accelerates as it changes direction going around a circle. The translational acceleration is given by the letter a. Rotational acceleration is given by the Greek letter alpha α. To convert translational acceleration a in $\frac{meters}{second^2}$ into rotational acceleration α in $\frac{radians}{second^s}$ use the equation $\alpha = \frac{a}{r}$.

Think of the ice skater who pulled her arms in in 1 second decreasing her radius of rotation, her velocity (v) of rotation increased, meaning that she rotated faster — accelerated. She rotated more times per second, so her **angular velocity** increased, which is **angular acceleration**. If she was rotating at 3 times per second, which increased to 10 times per second, her angular rotation went from $3(2\pi) \frac{radians}{second}$ to $10(2\pi) \frac{radians}{second}$ or $6\pi \frac{radians}{second}$ to $20\pi \frac{radians}{second}$.

$$\frac{20\pi \frac{radians}{second} - 6\pi \frac{radians}{second}}{second} = 14\pi \frac{radians}{second^2}$$

Translational motion is movement in a straight line and rotational motion is movement back to where it started in a circle. Just as you have velocity, acceleration, and momentum with translational motion, they can also be measured in rotational motion.

The momentum of an object moving in a straight line is conserved (unchanged) unless it is accelerated; likewise, angular momentum (in a circle) is conserved unless it is accelerated. When the radius of rotation is decreased (as when the arms are drawn in towards the body of an ice skater), the angular velocity (measured as rotations per second) increases because the rotation is completed in less time. Angular momentum is vr which is conserved, so as r decrease v must increase to keep vr constant.

LABORATORY 8

Rotational Motion

REQUIRED MATERIALS

- Piece of string 1.2 meters long
- Metric ruler or meter stick
- Dark marker
- 5 metal washers
- Paper or plastic straw
- Spring scale
- Eye hook
- Piece of wooden 2 × 4
- Stopwatch or timing app on smart phone
- Calculator
- Piece of string 1 meter long
- Tennis ball
- Tape

Introduction

Translational velocity is the velocity in $\frac{meters}{second}$ that is at a right angle (perpendicular) to the outer end of the radius of rotation. Rotational velocity is the number of times an object rotates or revolves in a minute or second. It can be in units of revolutions per minute (RPMs) or radians per second.

The circumference of a circle is found by $C = 2\pi r$.

The translational velocity is found by $v = \dfrac{C \times \frac{revolutions}{minute}}{60}$. Divide by 60 to get meters per second rather than meters per minute.

The rotational velocity is found by $w = \frac{v}{r}$.

Purpose

This exercise provides experience in measuring the radius and rotational velocity (RPM) of a revolving object. From these data the circumference of the circle, translational and rotational velocities ($\frac{radians}{second}$) can be calculated.

Procedure

 observe

1. Cut a piece of string 1.2 meters (1 meter with 10 centimeters extra on each end) long. With a dark marker, mark it 10 cm from each end. Place a dark mark every 10 cm along the 1 meter length of the string.

72 | Physics

2. Tie one end of the string to 5 metal washers.
3. Cut a paper or plastic straw to a length of 5 cm.
4. Thread the string through the straw.
5. Tie the loose end of the string to the hook on the end of a spring scale.
6. Screw an eye hook into a stable wooden surface (like a piece of 2×4).
7. Attach the other end of the spring scale to the eye hook.

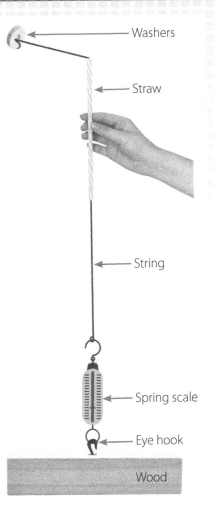

question

8. The spring scale gives the force that the string pulls away from the center of the circle, so it is the centrifugal force.
9. Carefully hold the section of the straw so that you can twirl the metal washers in a circle. Be carefully not to hit yourself or anyone else.
10. What happens when you spin the washers faster? This is an increase in the translational and rotational revolution.
11. What happens when you spin the washers more slowly? This is a decrease in the translational and rotational revolution.

research

12. Spin the washers so that the second 10 cm mark on the string is at the top of the straw. This gives a radius for the circle of revolution of 20 cm.
13. Practice this so that you can spin the washers steadily with this radius.
14. Have someone count the number of times the washers go around the circle in a minute.

hypothesis

15. Calculate the circumference of the circle. Notice that the radius is 0.2 m.
16. Calculate the translational velocity of the washers. Divide the revolutions per minute by 60 to get the revolutions per second. Do not worry if this turns out to be a decimal less than 1.
17. Calculate the rotational velocity in $\frac{radians}{second}$.

experiment

18. Repeat steps 12 through 17, except this time spin the washers so that the third cm mark on the string is at the top of the straw.

analyze

19. Do the results from steps 15 through 18 agree with your observations in steps 10 and 11?
20. Tape a tennis ball to the end of a 1 meter long piece of string.
21. Hold the other end of the string with your hand and twirl the string so that it winds around your arm. Notice the velocity of the tennis ball as it goes around getting closer to your arm. Record your results.

conclusion

22. Explain what happened with complete sentences.

CHAPTER NINE
PROJECTILE MOTION

OBJECTIVES

At the conclusion of this lesson the student should have an understanding of

- The kinetic and potential energies of an object thrown in the air
- The mechanics (motion) of a baseball in a major league home run
- What happens when you shoot an arrow at a distant target
- The escape velocities of spacecraft

74 | Physics

Note: There are a few more advanced formulas used in this chapter. We do try to make them as clear as possible in principle and solution.

If you threw a baseball straight up in the air, it would have maximum kinetic energy when it left your hand. As it went higher, its kinetic energy would decrease (because it is slowing down), and its potential energy would increase (because it is getting higher). When it reached its greatest height, it would have maximum potential energy and zero kinetic energy. As it fell back down, its potential energy would decrease, and its kinetic energy would increase. As it fell and reached the same height as when it left your hand, its potential energy would be zero and its kinetic energy would be the maximum.

A regulation major league baseball has a mass of 145 grams. If you threw the ball at an initial velocity of 15 $\frac{m}{s}$, what would be its initial kinetic energy?

$$KE = \tfrac{1}{2}mv^2 = \tfrac{1}{2}(0.145 \text{ kg})\left(15\tfrac{m}{s}\right)^2 = 16.3 \text{ J}$$

How high would the ball go? The acceleration of the ball is negative because it is slowing down, so the relationship between the distance (height), time and velocity is...

$$d = v_o t - \tfrac{1}{2}gt^2$$

where v_o is the initial velocity (at time zero) of the ball and t is the time it takes the ball to reach it maximum height. This equation will be used shortly. g is 9.8 $\frac{m}{s^2}$. Before you can find d, you must know (t) how long it takes the ball to get to the highest point. This relationship is provided for you. In a more advanced course, you would find out where it came from.

$$t = \frac{v_o}{g} = \frac{15\frac{m}{s}}{9.8\frac{m}{s^2}} = 1.53 \text{ s}$$

Take the average of the initial velocity and the final velocity ($0 \frac{m}{s}$) where the average velocity is $\frac{v_o + v_f}{2} = \frac{15\frac{m}{s} - 0\frac{m}{s}}{2} = 7.5 \frac{m}{s}$ where v_f is the final velocity at the ball's highest point.

$$d = \left(v_{average}\right)t = \left(7.5\tfrac{m}{s}\right)(1.53 \text{ s}) = 11.48 \text{ m}$$

What is the potential energy of the ball at its greatest height?

$$PE = mgh = (0.145 \text{ kg})\left(9.8\tfrac{m}{s^2}\right)(11.48 \text{ m}) = 16.3 \text{ J}$$

This is the same as the kinetic energy of the ball when it was first released from your hand. When the ball is going up, its

$$KE + PE = 16.3 \text{ J}$$

As the ball gets higher, its kinetic energy gets less because it is slowing down and as the ball is falling back down, its potential energy decreases because its height is decreasing, and its kinetic energy is increasing because it is going faster (*Diagram 9–1*).

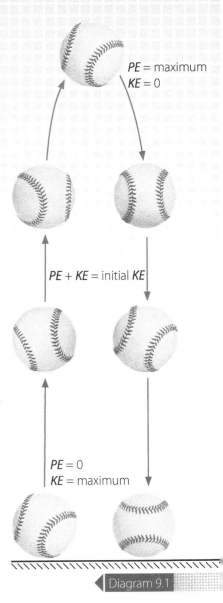

Diagram 9.1

KE = kinetic energy
m = mass
v = velocity
t = time
v_o = initial velocity
PE = potential energy

Recently a major league baseball player hit a home run that had an exit velocity of 106 $\frac{\text{miles}}{\text{hour}}$ at an angle of 41°.

If the ball went up at a 45° angle, it would go the greatest distance. So, 41° is very good. Assuming minimal air resistance and wind, how high and how far could the ball go?

The units must first be converted into metric units. Metrics are often used in physics because everything is a multiple of 10, which is much easier to work with.

There are approximately 1609 meters in a mile, so . . .

$$\left(106 \frac{\text{miles}}{\text{hour}}\right)\left(1609 \frac{\text{meters}}{\text{mile}}\right) = 170,554 \frac{\text{meters}}{\text{hour}}$$

There are 3600 seconds in an hour, so . . .

$$\frac{170,554 \frac{\text{meters}}{\text{hour}}}{3,600 \frac{\text{seconds}}{\text{hour}}} = 47 \frac{\text{meters}}{\text{second}}$$

The baseball has a curved path which introduces some complex math. A handy way to deal with it is to break it down into vertical motion and horizontal motion. A similar thing was done in the earlier chapter dealing with vectors. A rectangle is drawn with the diagonal being the trajectory of the ball. It is 47 mm (millimeters) long representing 47 $\frac{\text{meters}}{\text{second}}$.

The horizontal line (x–axis) is measured to be 36 mm long representing 36 $\frac{\text{meters}}{\text{second}}$. The vertical line is measured to be 32 mm long representing 32 $\frac{\text{meters}}{\text{second}}$ (Diagram 9.2).

Another way to do this is to use trigonometry. If you have not had trigonometry yet, here are some basics that you need for now. In the following diagram, the side (called y) opposite the angle divided by the hypotenuse (diagonal) is called the sine (abbreviated sin) of the angle. The side (called x) adjacent to the angle divided by the hypotenuse is called the cosine (abbreviated cos) of the angle (Diagram 9.3).

v_y stands for the initial vertical velocity and v_x stands for the initial horizontal velocity.

$$\frac{v_y}{47 \frac{m}{s}} = \text{sine } 41° = 0.656$$

$$v_y = \left(47 \frac{m}{s}\right) 0.656 = 30.8 \frac{m}{s}$$

To find the sine of 41°, select 41 on your calculator followed by the button that has "sin".

$$\frac{v_x}{47 \frac{m}{s}} = \text{cosine } 41° = 0.755$$

$$v_x = \left(47 \frac{m}{s}\right) 0.755 = 35.5 \frac{m}{s}$$

To find the cosine of 41°, push 41 on your calculator followed by the button that has "cos."

Notice that by using the vector diagrams, $v_y = 32 \frac{m}{s}$ and $v_x = 36 \frac{m}{s}$.

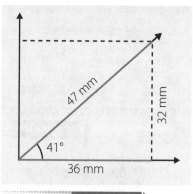

Diagram 9.2

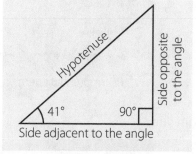

Diagram 9.3

For the vertical velocity, $32 \frac{m}{s}$ is close to $30.8 \frac{m}{s}$, and for the horizontal velocity, $36 \frac{m}{s}$ is $35.5 \frac{m}{s}$ rounded off. When it is important to have more accurate numbers, the trigonometry method is better, though it is a bit beyond the scope of this course, except as a reference.

To find how long it took the baseball to reach its greatest height this relationship is used.

$$t = \frac{v_0}{g} = \frac{30.8 \frac{m}{s}}{9.8 \frac{m}{s^2}} = 3.14 \text{ s}$$

v_0 means the initial velocity or the velocity at time zero. It would take the ball 3.14 seconds to reach its highest height. If the baseball came back down to the height at which it was hit by the bat, the time would be twice this or 6.28 s. But, as you know, the ball never comes back to its original height because it lands in someone's glove up in the stands.

The highest point that the ball reached before coming back down was . . .

$$d_y = v_0 t - gt^2 = \left(30.8 \frac{m}{s}\right)(3.14 \text{ s}) - \frac{1}{2}\left(9.8 \frac{m}{s^2}\right)(3.14 \text{ s})^2 = 96.7 \text{ m} - 48.3 \text{ m} = 48.4 \text{ m}$$

There are 0.305 meters in 1 foot, so . . .

$$\frac{48.4 \text{ m}}{0.305 \frac{m}{foot}} = 159 \text{ feet}$$

How far horizontally would the ball have gone if it had not landed in the stands? The total time the ball would have been in the air is 6.28 s. The initial force of the bat on the ball would have been the only force acting on the horizontal part of the ball's motion because gravity is considered in the vertical motion. Without air resistance or wind, the ball would have kept going horizontally according to Newton's first law of motion (an object in motion will stay in motion unless there is an opposing force). So, the horizontal distance is the velocity times the time.

$$d_x = \left(35.5 \frac{m}{s}\right)(6.28 \text{ s}) = 223 \text{ meters}$$

$$\frac{223 \text{ meters}}{0.305 \frac{m}{foot}} = 731 \text{ feet}$$

The ball does not really make it that far because it lands up in the stands where a fortunate fan goes home with a souvenir.

If the ball went up at a 45° angle . . .

$$v_y = \left(47 \frac{m}{s}\right)(\sin 45°) = (47)(0.707) = 33.2 \frac{m}{s}$$

$$v_x = \left(47 \frac{m}{s}\right)(\cos 45°) = (47)(0.707) = 33.2 \frac{m}{s}$$

$$t = \frac{v_0}{g} = \frac{33.2 \frac{m}{s}}{9.8 \frac{m}{s^2}} = 3.39 \text{ s}$$

$$d_y = v_0 t - \frac{1}{2}gt^2 = \left(33.2 \frac{m}{s}\right)(3.39 \text{ s}) - \frac{1}{2}\left(9.8 \frac{m}{s^2}\right)(3.39 \text{ s})^2 = 112.5 \text{ m} - 56.3 \text{ m} = 56.2 \text{ m}$$

$$\frac{56.2 \text{ m}}{0.305 \frac{m}{foot}} = 184 \text{ feet}$$

$$d_x = v_x(2t) = \left(33.2 \frac{m}{s}\right)(6.78 \text{ s}) = 225 \text{ m}$$

$$\frac{225 \text{ m}}{0.305 \frac{m}{foot}} = 738 \text{ feet}$$

This is greater than the 731 feet when the ball went up at a 41° angle.

$d_y =$
$v_0 =$ initial velocity
$t =$ time
$d_x =$
$v_y =$ vertical velocity
$v_x =$ horizontal velocity

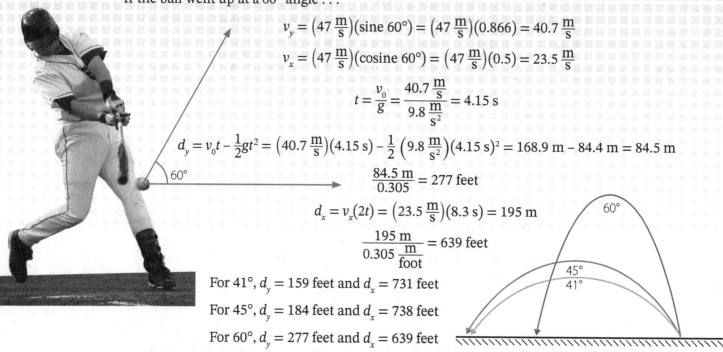

If the ball went up at a 60° angle . . .

$v_y = \left(47 \frac{m}{s}\right)(\sine 60°) = \left(47 \frac{m}{s}\right)(0.866) = 40.7 \frac{m}{s}$

$v_x = \left(47 \frac{m}{s}\right)(\cosine 60°) = \left(47 \frac{m}{s}\right)(0.5) = 23.5 \frac{m}{s}$

$t = \frac{v_0}{g} = \frac{40.7 \frac{m}{s}}{9.8 \frac{m}{s^2}} = 4.15 \text{ s}$

$d_y = v_0 t - \frac{1}{2}gt^2 = \left(40.7 \frac{m}{s}\right)(4.15 \text{ s}) - \frac{1}{2}\left(9.8 \frac{m}{s^2}\right)(4.15 \text{ s})^2 = 168.9 \text{ m} - 84.4 \text{ m} = 84.5 \text{ m}$

$\frac{84.5 \text{ m}}{0.305} = 277$ feet

$d_x = v_x(2t) = \left(23.5 \frac{m}{s}\right)(8.3 \text{ s}) = 195 \text{ m}$

$\frac{195 \text{ m}}{0.305 \frac{m}{foot}} = 639$ feet

For 41°, $d_y = 159$ feet and $d_x = 731$ feet

For 45°, $d_y = 184$ feet and $d_x = 738$ feet

For 60°, $d_y = 277$ feet and $d_x = 639$ feet

As the angle gets greater, the ball goes higher. A 45° angle gives the greatest horizontal distance. Angles above and below 45° give shorter horizontal distances. In hitting an out-of-the-ballpark home run, the ball must go high and far enough to clear the back wall of the stadium. These distances vary a bit between different stadiums. Wind and humidity also have bearing that can vary quite a bit. Major league baseball players are not ignorant. They are knowledgeable and apply these principles of physics in their practice and performance.

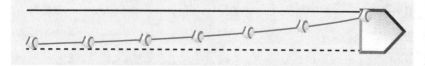

Path of a curve ball

Sometimes a good pitcher puts added motion on the ball — such as in a curveball where the ball curves away from the batter. Looking down from above, the pitcher is observed to spin the ball in a counterclockwise direction before release when throwing to a right-handed batter. While the ball is spinning, its motion through the air and the roughness of the ball causes air to go around the ball faster on the left side than on the right side. This produces a slight vacuum effect, causing the ball to move to the left away from the batter. Some debate about whether the seams on the ball influence the air flow around the ball. The ball moving through the air has the same effect as if the air were moving past the ball.

If you shot an arrow at a target, the arrow would be falling under the influence of gravity as well as moving toward the target. So, if you aim directly at the target, the arrow may hit below the target. You would have to compensate by aiming a little higher than the target.

If you shot an arrow at an apple just as the apple came loose from a tree, the arrow and the apple would fall at the same rate and you would hit it if you aimed directly at it.

A spacecraft launched from earth can be thought of as a projectile as well. As a spacecraft gets farther from earth, the gravitational attraction decreases by the inverse square law that we looked at earlier. If the spacecraft got to a height of 10 times the radius of earth, the gravitational attraction would be one hundredth ($\frac{1}{100}$) — 10^2 is 100 and $\frac{1}{100}$ is the inverse of 100. There is a critical launch velocity that would place the spacecraft far enough away with enough remaining velocity to escape earth and continue traveling. The higher the spacecraft gets, the more its velocity decreases. Think of the example of the ball thrown in the air and coming back down. The **escape velocity** depends upon the gravitational constant, the mass of the attracting body (such as earth), and the distance at launch from the center of the attracting body.

$$v^2 = G\frac{m}{d}$$

So, the escape velocity is the square root of the gravitational constant times the mass of the attracting body divided by the distance from the center of the attracting body at launch. A spacecraft needs an escape velocity of 11 $\frac{Km}{s}$ to escape earth's gravity.

The sun has a mass of 330,000 times that of earth and a radius (distance from its center) of 109 times that of earth. A spacecraft launched from the sun would have to overcome an escape velocity of 620 $\frac{Km}{s}$. That could be tricky to launch from the sun. Mars has an escape velocity of 10.4 $\frac{Km}{s}$ because it is just a little smaller than earth. The moon has an escape velocity of 2.4 $\frac{Km}{s}$ because it is smaller than earth. The word Moon is capitalized when used as a proper name. A satellite is called a moon with a lower case "m." Our moon is not a satellite because it is far too big compared to earth to be a satellite of earth. Notice that in every case, the escape velocity depends upon the size of the attracting body, not that of the spacecraft.

The trajectory of a baseball is very simple compared to that of spacecraft sent beyond earth. The most complex and enlightening space exploration undertaken was that of *Voyagers 1* and *2*. This was called "The Grandest Tour." More was learned about the solar system (the system around the sun) than all the other space explorations put together. Astronomy curricula and textbooks all had to be rewritten. Every planet and every moon around each planet are very different from each other. The uniformity expected from an evolutionary model was not there. The heavens declared the glory of God most definitely. A principle that I have seen consistently over the years is that good, honest scientific research will ultimately confirm God's testimony and creative role. Many will have a biased interpretation of data that dismisses God's role. But remember, that is the interpretation not the data.

NASA *Voyager* spacecraft

Voyager 1 was launched on September 5, 1977, and explored Jupiter and its moons and Saturn and its rings and moons. It changed trajectory in 1980 to go perpendicular to the plane of the solar system and leave the solar system. It used the gravitational fields of the planets to increase its velocity (boomerang effect) to leave the Solar System.

Voyager 2 was launched August 20, 1977, 16 days before *Voyager 1*. It had a different trajectory than *Voyager 1*, which had a shorter and faster path. It also explored the systems of Jupiter and Saturn. It went beyond to explore the systems of Uranus and Neptune. It found some unexpected, unusual sights such as shepherding satellites of Saturn that braided some of the rings of Saturn and rings around Uranus and Neptune. It too achieved the escape velocity to leave the solar system. Both *Voyagers* are currently exploring interstellar (between star systems) space and sending data back to earth.

Preparing the trajectories of the *Voyagers* took some heavy-duty calculus. This is involved in the field called computational physics.

In summary, as a projectile goes higher, it slows down and its kinetic energy decreases while its potential energy increases. When it falls back down (it goes faster and its kinetic energy increases) and its potential energy decreases.

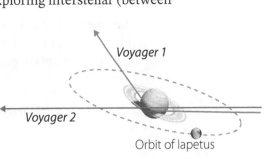

Voyager 1
Voyager 2
Orbit of Iapetus

LABORATORY 9

Projectile Motion

REQUIRED MATERIALS

- 2 marbles
- Metric ruler
- Stopwatch or timer app on smart phone
- Tennis ball or baseball
- Protractor to measure angles
- **Note:** An assistant is needed to help with this lab.

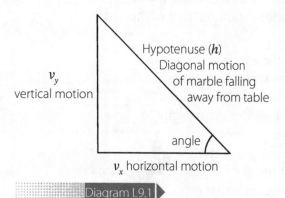

Diagram L9.1

Introduction
The motion of a projectile can be broken down into vertical and horizontal motions using a vector diagram where the vertical line represents the vertical motion and the horizontal line represents the horizontal motion (*Diagram L9.1*).

The vertical motion can also be found by multiplying the diagonal (combined motion) by the sine of the angle of the projectile pathway. The horizontal motion can also be found by multiplying the diagonal (combined motion) by the cosine of the angle of the projectile pathway.

Purpose
This exercise demonstrates how the vertical and horizontal components of projectile motion can be determined for observed projectiles.

Procedure

 observe

1. Place 2 marbles on the edge of a tabletop. Simultaneously cause one marble to fall over the side of the table and push the second one so that it falls away from the table. It is important that you do your best to cause them both to fall at the same time. Have someone else time how long it takes the marbles to reach the floor. One marble should land right next to the table. It has a vertical distance but no horizontal distance. The second marble will fall the same distance but has a horizontal distance unlike the first marble. If the marbles do not fall at the same time, do not use their time. Measure the distance that the marbles fell from the table to the floor. Measure how far away from the table the second marble fell.

2. v_y = the distance the marbles fell divided by the time.

3. v_x = the distance that the second marble fell away from the first marble divided by the time.

4. Draw a vector diagram like the one below to find the combined velocity of the second marble. The diagonal represents the combined vector. The marble fell in a curved pathway, but we are using a straight line (which is close to the actual pathway) to keep the math manageable (*Diagram L9.2*)

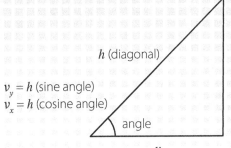

$v_y = h$ (sine angle)
$v_x = h$ (cosine angle)

Diagram L9.2

question

5. Measure the length of the diagonal. What combined velocity does it represent?

research

6. Using a protractor, measure from your vector diagram the angle between the diagonal and the top line. Use your calculator to find the sine and cosine of this angle.

hypothesis

7. Calculate the length of the diagonal times the sine of the angle. How does this value compare to the value given by the length of the vertical line?

8. Calculate the length of the diagonal times the cosine of the angle. How does this value compare with the value given by the length of the horizontal line?

experiment

9. Go outside and throw a baseball or tennis ball. Straight up is 90°, horizontal is 0°, and 45° is in between 90° and 0°. Try to throw the ball at 45°. Then throw the ball at less than 45° with the same velocity.

analyze

10. How far did it go compared to the 45° angle? Throw the ball at greater than 45° with the same velocity. How far did it go compared to the 45° angle?

conclusion

11. Do these results seem reasonable based upon what you learned in this lesson? Explain.

Laboratory 9 | 81

CHAPTER TEN

KEPLER'S LAWS OF PLANETARY MOTION

OBJECTIVES

At the conclusion of this lesson the student should have an understanding of

- The nature of science and revelation
- Observations of stars and planets
- Early heliocentric model of Aristarchus of Samos
- Geocentric models of Aristotle and Ptolemy
- Heliocentric model of Copernicus
- Tycho Brahe's observations
- Kepler's Laws of Planetary Motion
- The contributions of Galileo and Newton accounting for the centripetal acceleration that keeps planets in orbit around the sun

The first nine chapters dealt with the area of physics called **mechanics**, which is the study of motion. This chapter is an application of those concepts. An important aspect of science is the purpose and limitation of science. This is not meant to be a negative statement. Consider all the blessings that we have gained from the many long years and hours that thousands of people have put in so that we could enjoy these blessings. The laptop that I am writing on is one of those blessings. I remember when I used a typewriter and could not just back up and correct it when I made a mistake. I could not go back and insert a sentence that I left out. Science, by nature, grows and never arrives at ultimate truth. That is what this chapter is all about. In Scripture, we have the inspiration of the Holy Spirit to give us truth in written form. Scripture has many prophecies that have been fulfilled later to the letter. We cannot do that in science. We can make reasonable conclusions — such as if you hit your thumb with a hammer, it is going to hurt. Prophecies are different. Prophets told many years before precisely where Christ would be born. Even the Roman officials that occupied Israel at the time recognized that. Science is a different type of tool than Scripture which is the inerrant Word of God. Each has its purpose and when you use the wrong tool for a task you have problems.

The heavens have always been a subject of curiosity. We are told that on the fourth day of creation "God made the two great lights; the greater light to rule the day, and the lesser light to rule the night. He made the stars also" (Genesis 1:16).

Stars and planets have always fascinated people. The Chaldeans, from which God called Abraham, named some of the prominent constellations. These are regions of the night sky recognized by groupings of brighter stars. God used stars in His promise to Abraham.

> Then He brought him outside and said, "Look now toward heaven, and count the stars if you are able to number them." And He said to him, "So shall your descendants be." And he believed in the Lord, and He accounted it to him for righteousness (Genesis 15:5–6).

Even though Abraham could only see the closest stars in our Milky Way **galaxy**, there were so many that Abraham could not count them all.

The word *planet* comes from the word "wanderer" in Greek. They noticed that stars always appeared grouped in the same constellations over the course of a human life span. They noticed other bright objects in the sky that appeared at different times in different constellations, so they called them wanderers in contrast to the stars. In Jude 13, the Greek word for wanderer was used for false teachers.

Throughout this chapter there are names of scientists with the years they lived or wrote. These dates are not given to be memorized but to show which ideas were developed before others and which were contemporaries and could have shared ideas with each other.

In 300 B.C., Aristarchus of Samos said that the sun was the center of the universe. This meant that all the heavenly bodies, including the planets, revolved around the sun. Even though we know today that the sun is not in the center of the universe, we know that planets revolve around the sun in our solar system. This idea was lost by the Greeks and only realized much later.

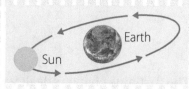

Geocentric model

Aristotle (384–322 B.C.) and Plato taught that earth was surrounded by four spheres. On the outer sphere were all the stars in their constellations. As this sphere rotated around earth, the stars appear to pass overhead across the night sky. Planets were on a smaller sphere inside the sphere of the stars. This was their explanation for why the planets appeared to wander apart from the stars. The sun and moon were also on separate spheres revolving around the earth. They thought of the spheres as ideal shapes appropriate to the heavens, as approaching perfection because they were closer to God.

In A.D. 150, Ptolemy published a work called the Almagest (The Greatest) where he described the earth as being the center of the solar system with the planets, sun, and moon revolving around the earth. This was a modified form of the teachings of the Greeks. Popular ideas at that time were common sense, just as they are today. This, along with the idea that earth was the center of the solar system, became part of the teachings of the Roman Catholic Church at that time. Later, in the time of the Renaissance, Reformation, and the Inquisition, to think otherwise was heresy and subject to the death penalty.

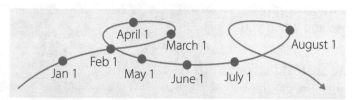

Diagram 10.1
Retrograde motion of an outer planet

The **geocentric** (earth-centered) idea proposed by Ptolemy did a pretty good job of predicting where and when the planets would appear in the night sky. There were some problems, however, including the changing brightness of planets and the retrograde motion of Mars, Jupiter, and Saturn. They had not discovered Uranus, Neptune, and Pluto yet. Retrograde motion means to go backward. As these planets move across the night sky on successive nights, they appear to go backward and then forward again (*Diagram 10.1*).

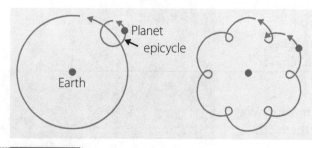

Ptolemy tried to explain this by having the planets go in smaller circles (called epicycles) as they gradually revolved around earth. This would make them appear to go backward and then forward again. But this still did not explain the changes in the brightness of the planets (*Diagram 10.2*).

Notice, as in this case, in the development of ideas in science, observations are made and must be accounted for in the proposed explanations.

Diagram 10.2
Orbit of a planet with epicycles

Much later, the Polish astronomer Nicolaus Copernicus (1473–1543) published, while on his deathbed, a **heliocentric** (sun-centered) view. This was a radical departure from the long-held view described earlier by Ptolemy. He waited until they could not do anything to him, because others had been burned at the stake for lesser heresies.

Copernicus' ability to predict the positions of the planets in the night sky was about the same as that of Ptolemy. But he explained the retrograde motion of planets in a much simpler way. With the planets revolving around the sun, the planets closer to the sun would not have to go as far as the planets

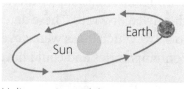

Heliocentric model

farther away, so they would pass the outer planets and go around the sun sooner, making it appear that the outer planets went backward and forward again (*Diagram 10.3*).

He explained the changing brightness of the planets, because if they all revolved around the sun, sometimes they would be closer to earth and sometimes farther away.

Some argued against Copernicus' ideas stating that if earth revolved around the sun, it would be going about 1,000 miles an hour and we would feel it and blow off the earth because of the great wind it would produce. This was concluded because of their lack of understanding of gravity that came later with Galileo and Newton. It was later realized that we did not pass through air, but that our atmosphere was held to earth by gravity, so it traveled with us.

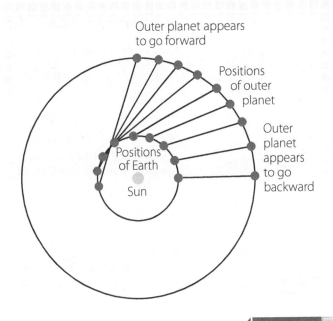

Diagram 10.3

Retrograde motion of outer planet with heliocentric model

Much later the heliocentric model became more acceptable. Usually it takes about 100 years for a completely different idea to replace a common belief.

It is common today in astronomy classes to compare the church's treatment of Copernicus to the treatment of evolutionists today. The point is made that church dogma hindered reliable scientific observations and sound conclusions. They think of creation as church dogma and evolution as conclusions made from reliable observations. If doctrines, such as creation, were just opinions of church officials and not divine revelation, their claims would be valid. The problem is how the Scriptures are viewed. You more than likely will encounter such arguments. When you do, ask how they view the Scriptures. Many view creation as an idea where there are no changes at all in living forms. But creation acknowledges that limited changes do occur in living organisms that are not evolution, because they are restricted within the original kinds of creation and are controlled by DNA that had to be present in first life forms created.

Further detailed observations of the positions of stars and planets were made by the Danish astronomer Tycho Brahe (1546–1601) over the span of many years. As a teenager, Tycho Brahe noticed that the astronomers of his day (without telescopes, because they had not yet been invented) predicted the time and day that a total eclipse of the sun (the moon coming between the earth and the sun, making the sun go dark momentarily) and it occurred just as they predicted. But he also noticed that some of their other predictions did not come out so well. He developed a passion to make better predictions and, thereby, have a better understanding of the universe. As he became a young man, he had a bit of a combative spirit. Once, he got into a duel and got his nose blown off with a musket ball. The doctor made him a replacement nose out of brass. The king of Denmark, wanting to protect him, built him an observatory on an island off the coast of Denmark so that he would stay there out of harm's way. Brahe built a large instrument, called a quadrant, that look like half of a giant protractor. By looking past the quadrant with its markings to distant objects in the sky, he

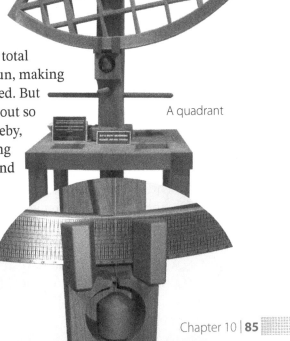

A quadrant

could accurately identify their positions within $\frac{1}{60}$ of a degree. The sky from horizon to horizon is 180 degrees and $\frac{1}{60}$ of a degree is called a minute. This enabled him to make very accurate observations.

Even though he kept very accurate records over many years, he was not well organized. He did not have a spread sheet to record his data. He had a model of the solar system where the planets revolved around the sun, but he had the sun and moon revolving around earth. To make better predictions, he had to be able to see general patterns from all his data. Johanne Kepler succeeded Brahe and was able to see several patterns from his data which became known as Kepler's Laws of Planetary Motion. At first, Kepler had the same limitations in predicting the positions of the planets as Ptolemy and Copernicus because he viewed the orbits of planets as being circles. Tradition has it that one day he said, "What a silly bird I have been." He tried placing the orbits of the planets in ellipses, instead of circles, and his predictions were much better. An ellipse has two points around which an object revolves instead of one as in a circle (Diagram 10.4).

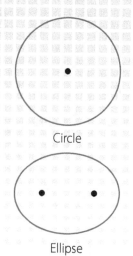

Circle

Ellipse

Diagram 10.4

Kepler's First Law of Planetary Motion was that planets orbit around the sun in ellipses instead of circles. It was a minor but significant improvement because the ellipses were almost circles.

Position of Earth about the sun

Each of these lines is the radius vector

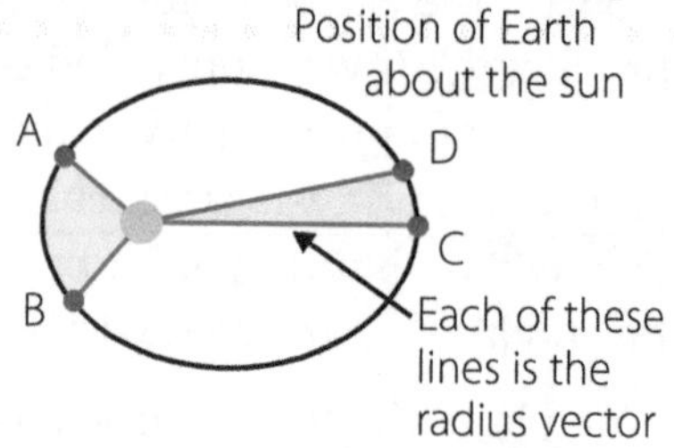

The two shaded regions are equal in area. Earth goes from A → B in the same time it goes from C → D

Diagram 10.5

The Second Law of Planetary Motion was that planets go faster when they are closer to the sun and slower when they are farther from the Sun. Another way to state it is that the radius vector from the sun to a planet sweeps out equal areas in equal amounts of time. It is shown by this diagram (Diagram 10.5).

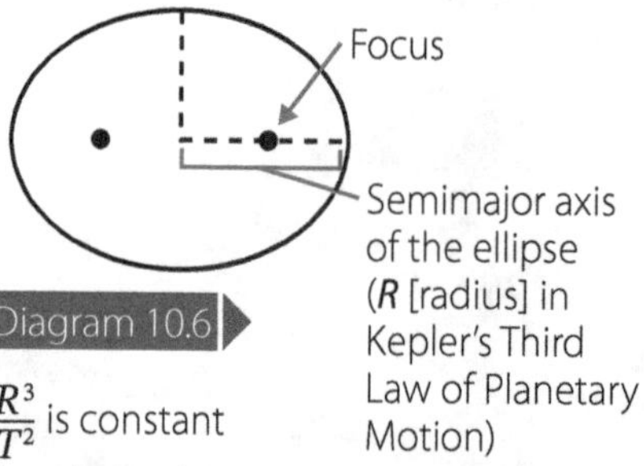

Focus

Semimajor axis of the ellipse (R [radius] in Kepler's Third Law of Planetary Motion)

Diagram 10.6

$\frac{R^3}{T^2}$ is constant for planets in the solar system

The Third Law of Planetary Motion, sometimes called the law of harmonies, is that the radius of revolution of a planet cubed, divided by the period (T, the time it takes a planet the orbit the Sun) squared is the same for every planet except for Pluto (that had not been discovered yet) (Diagram 10.6).

These three laws of planetary motion together were much better at predicting the positions of planets than before. Tycho Brahe would have been thrilled if he could have lived to see the fruition of his work. Accurate predictions are evidence of design and numerical predictions are much greater evidence of design. Kepler gave God the glory for His creation of the heavens. Kepler had the earth revolving around the sun, and the moon revolving around earth.

In science, everyone's work is built upon the work of many others that have gone before. There will always be some unanswered questions. When planetary orbital patterns were better understood, the question that remained was, why do the planets keep changing direction to follow around the sun? What keeps them going around the sun? Why don't they just fly off into space? In chapter 8,

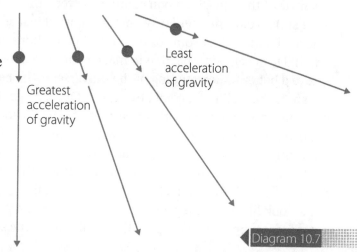

Greatest acceleration of gravity

Least acceleration of gravity

Diagram 10.7

86 | Physics

The Galilean moons Io, Europa, Ganymede, and Callisto (in order of increasing distance from Jupiter)

we called this **centripetal acceleration**. Acceleration is a change in velocity, and planets must constantly change direction to keep going around the sun. Remember that centripetal means "center seeking." But what was the force causing the centripetal acceleration? Whenever you have an acceleration, it must be caused by a force. This led later for Galileo (1564–1642) to propose the idea of the acceleration of gravity. He demonstrated, using inclined planes, that the greater the slant of the plane, the less the acceleration. This meant that the greatest acceleration was straight down (*Diagram 10.7*).

Many believe that answers to questions like this mean that nature can function without God having to make it work. This is based on the belief that God is only involved when He must be — when there is no explanation other than a miracle. This is contrary to a biblical understanding of the very nature of God. Maybe that is why we are tempted to act as if God does not know what we are thinking and doing.

Galileo was the first to report observing the moon with its craters, sunspots, and four of the moons around Jupiter that became known as the Galilean moons. He was one of the first to use a refracting telescope, one that uses lenses. He ended up living out the rest of his life in house arrest because the round heavenly bodies represented steps toward holiness to many of the religious leaders of his day. Sunspots and moon craters were thought of as blemishes in what was considered holy. This illustrates how ideas that are not spelled out in Scripture can lead to very serious problems.

Later Isaac Newton (1642–1727) proposed the Universal Law of Gravity that provided the force for the centripetal acceleration of the planets. He said that if you launched a projectile from a high enough mountain fast enough, it would orbit the earth. He stated that gravity was the force that provided the centripetal acceleration that kept the planets in orbit.

It is fascinating that just over two centuries after Newton wrote of a projectile being launched into space and orbiting the earth in his work *A Treatise of the System of the World*, the first such rocket fulfilled his thought experiment. In October 4th of 1957, the USSR launched Sputnik 1, an unmanned craft that orbited Earth until it fell back down in January 4th of 1958. The first person in space was launched into a single orbit of Earth in 1931, also from the USSR, and the first people to go to the moon came from the United States just a few years after that, in 1969. The race for space was accelerating.

Kepler's laws of planetary motion describe and explain the shape and times of planetary revolutions around the Sun with the heliocentric model.

LABORATORY 10

Elliptical Orbits And Moon Phases

REQUIRED MATERIALS

- $8\frac{1}{2}" \times 11"$ piece of cardboard
- Three $8\frac{1}{2}" \times 11"$ pieces of plain paper
- 2 tacks
- 10" long piece of string
- 2 helpers, a flashlight, and a darkened room

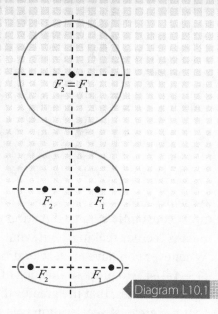

Diagram L10.1

Introduction

Planets and the moon reflect light from the sun and do not emit their own light. As the moon revolves around earth, the side illuminated by the sun is seen as bright. During a full moon, the earth is between the moon and the sun and we see light reflected to us showing the full surface of the moon. During a new moon, the moon is between us and the sun and the light from the sun hitting the moon reflects to the sun and we do not see it. During a first quarter moon (when the moon is a quarter of the way around earth), we see the right side of the moon illuminated because that is the side of the moon that the sun shines on. During a third (or last) quarter moon (when the moon is three quarters of the way around earth), we see the left side of the moon illuminated because that is the side of the moon the sun is shining on. As the moon revolves around earth, we are seeing light reflected from the part of the moon the sun shines on. While this is happening, the same side of the moon always faces us. This is because the moon revolves around earth at precisely the same rate that it rotates on its axis. If the rate of revolution and rotation were even slightly different, after a few thousand years we would be seeing a different part of the moon (*Diagram L10.2*).

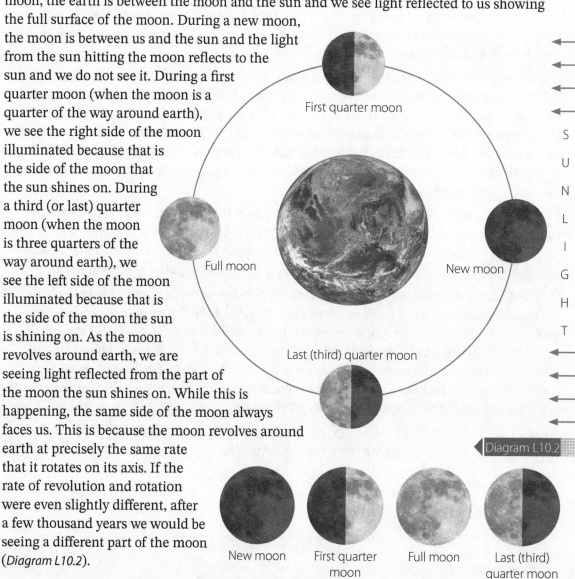

Diagram L10.2

88 | Physics

Purpose

This exercise is to demonstrate the nature of an elliptical orbit and the phases of the moon.

Procedure

 observe

1. Cut an $8\frac{1}{2}$" × 11" piece of cardboard or stiff paper. Lay a piece of unlined paper onto the cardboard. Cut a piece of string 10" long. Place a tack in the middle of the paper. Tie the two ends of the string together to make a loop. Place the string loop over the tack and a pencil in the other end of the string loop. With the string loop attached to the tack, draw a circle around the tack by moving it around the tack.

2. Place another sheet of paper on the cardboard. Instead of the tack, place 2 tacks 1 inch apart in the middle of the paper. Place the string loop over the two tacks and draw a small ellipse around the tacks. This is like a planetary orbit around the sun, where the sun is represented by one of the tacks. Notice that your ellipse has a point closer to the "sun" and a point farther from it (*Diagram L10.3*).

 question

3. Repeat step 2 except this time place the tacks 3 inches from each other. Draw out the ellipse around these foci. Notice that this a much narrower ellipse.

4. If you can, go outside tonight and see if you can see the moon. If the sky is relatively clear and the moon has not risen yet, check every night for a while until it is visible. As an alternative, you can go online and ask the question "What does the moon look like tonight?"

 research

5. Describe and draw the moon as you see it. After a week, go out and describe and draw the moon again. Do this 2 more times. Describe the phase of the moon for each evening that you observe it. This part of the lab will take at least a month to complete.

 hypothesis

6. The moon has phases which is evidence that it revolves around the earth. You can also see phases of the planet Venus if you use binoculars or a telescope. This is because Venus revolves around the sun closer to the sun.

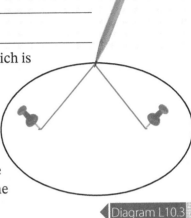

◁ *Diagram L10.3*

 experiment

7. Go into a dark room with 2 helpers. Stand in the middle of the room (you are earth). Have another person (the moon) walk in a circle around you, rotating so that the front of the person always faces you. The third person (the sun) is to stay on one side of the room and shine a flashlight on the person going around you. It is important that the light always come from the same direction. Have the rotating person go to new moon, first quarter moon, full moon, and third quarter moon position.

 analyze

8. Describe how the light shines on the rotating person at each position.

conclusion

9. This is how the phases of the moon are produced.

Laboratory 10 | 89

CHAPTER ELEVEN

HEAT ENERGY

OBJECTIVES

At the conclusion of this lesson the student should have an understanding of

- Thermal energy as the kinetic energy of atoms and molecules
- Fahrenheit and Celsius temperature scales
- Thermal equilibrium
- Calories, joules, and BTUs as units of heat energy
- Specific heat capacity
- Phase changes
- Heat transfer
- The greenhouse effect
- Relative humidity

Energy causes things to move. Within all substances, atoms and molecules are constantly moving, vibrating, and twisting due to their thermal energy. In 1857, Rudolf Clausius described heat as the kinetic energy of atoms and molecules.

If you keep striking a piece of metal with a hammer it will get warmer. The atoms in the metal are moved around and move faster. If the metal sits for a short time, it will cool back down. When something feels warm to the touch, we say that it has heat energy. When a piece of wood burns, oxygen atoms remove electrons that hold atoms of wood together. As the atoms of wood come apart, ashes are formed. In the winter when a pond freezes and you skate on it, the water molecules lose heat energy and slow down, forming ice.

Hot or cold are very subjective feelings. In a room full of people, you never ask if you should turn up the heat. Half the people will want it turned up and the other half will want it turned down.

Temperature gives us a more objective means of measuring heat, which is usually measured with a thermometer. In the early 1700s the German scientist Gabriel Fahrenheit developed an early thermometer consisting of a glass tube containing mercury. As it got colder, the mercury contracted and shrank down the tube into a reservoir at the bottom. As it warmed, the mercury expanded and went up the tube. He placed markings on the tube such that the freezing point of water was set at 32° and the boiling point of water was set at 212°. The Swedish astronomer Anders Celsius devised a thermometer in 1743 that had 0° for the freezing point of water and 100° for the boiling point of water. He divided his thermometer into 100 equal parts between 0° and 100°. The difference was that Celsius used the metric system. The word centigrade means divided into 100 parts, like the word centimeter. So, the proper name for his thermometer was the Celsius thermometer.

If you need to use both scales, which is rare, there are thermometers with both scales so you can read degrees Fahrenheit and degrees Celsius. There are 180 divisions between 32°F and 212°F on the Fahrenheit scale. There are 100 divisions between 0°C and 100°C on the Celsius scale — $\frac{180}{100}$ is $\frac{9}{5}$, so the ratio of Fahrenheit degrees over Celsius degrees in $\frac{9}{5}$. The Fahrenheit thermometer has the freezing point of water at 32°F while the Celsius thermometer has 0°C. To convert Celsius to Fahrenheit, take $\frac{9}{5}$ of the Celsius temperature and add 32.

$$T_f = \frac{9}{5}T_c + 32$$

To convert Fahrenheit to Celsius, just go back.

$$T_c = \frac{5}{9}(T_f - 32)$$

What is 20°C on the Fahrenheit scale?

$$T_f = \frac{9}{5}(20°C) + 32 = 36 + 32 = 68°F$$

What is 75°F on the Celsius scale?

$$T_c = \frac{5}{9}(75 - 32) = \frac{5}{9}(43) = 24°C$$

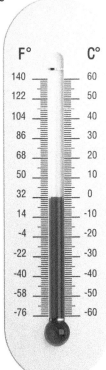

Because of the toxicity of mercury, many thermometers today contain colored alcohol.

Notice that the measurement of heat took place before (1700s) a better understanding of heat was proposed by Clausius in the 1800s. This is not uncommon in science. Being able to measure something often helps to better understand it.

In residential heating and cooling applications, the British unit of heat energy (BTU, British thermal unit) is often used — 1 BTU is equivalent to 0.252 Calorie.

Heat can also be understood as energy that is transferred from a hotter object to a colder object. You have experienced this many times. If the molecules of a hotter object, which are moving faster, collide with slower-moving molecules of a colder object, some of the slower molecules go faster and some of the faster molecules go more slowly. The hotter object became cooler (its molecules moved more slowly) and the colder object became warmer (its molecules moved faster). Eventually, the two objects will come to thermal equilibrium and be at the same temperature.

Different substances have different capacities to gain heat. Water can gain a great deal of heat with minimum temperature gain. This is because atoms in H_2O can use much of the energy by as internal motion in the H_2O molecule. Only when entire H_2O molecules move faster does the temperature go up. This property is called the specific heat capacity. The specific heat capacity of distilled water (with nothing dissolved in it) is 1 Calorie per °C per liter. This means that when 1 liter of water absorbs 1 Calorie of heat energy, it goes up 1°C. One liter of water is equivalent to 1 kg (kilogram) of water. A Calorie is used for the specific heat capacity of a kg of water and a calorie is used for the specific heat capacity of a gram of water — 1 Calorie (capital C) is 1,000 calories (lower case c). In biological and medical applications, the Calorie is the standard unit of heat energy). In physics, the joule is the standard unit of energy (including heat energy) where 1 Calorie is equivalent to 4148 J. In many applications today, the joule is replacing the Calorie.

How many Calories would it take to raise the temperature of 1 liter of water from 15°C to 20°C?

It takes 1 Calorie to raise 1 liter of water 1°C.

$$\Delta T = 20°C - 15°C = 5°C$$

It would take 5 Calories to raise 1 liter of water from 15°C to 20°C.

What is this in joules?

$$(5 \text{ Calories})\left(4148 \frac{J}{\text{Calorie}}\right) = 20{,}740 \text{ J}$$

How many Calories would it take to raise 1 liter of water at 72°F to 100°F?

First you must convert degrees F to degrees C.

$$T_c = \frac{5}{9}(72 - 32) = \frac{5}{9}(40) = 22°C$$

$$T_c = \frac{5}{9}(100 - 32) = \frac{5}{9}(68) = 38°C$$

$$\Delta T = 38°C - 22°C = 16°C$$

$$\frac{1 \text{ Cal}}{°C \text{ liter}}(16°C) = 16 \text{ Calories}$$

How many Calories would it take for 5 liters of water?

$$5 \text{ liters} \left(\frac{16 \text{ Calories}}{1 \text{ liter}}\right) = 80 \text{ Calories}$$

What is this in joules?

$$80 \text{ Calories} \left(4{,}148 \frac{J}{\text{Calorie}}\right) = 331{,}840 \text{ J} = 331.84 \text{ kJ (kiloJoule)}$$

Some other examples of specific heat capacities are...

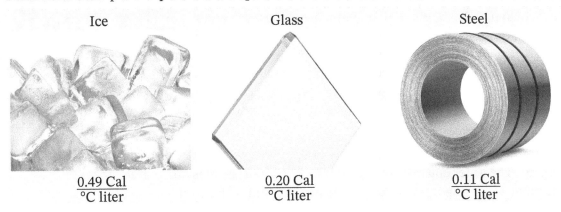

Ice
$$\frac{0.49 \text{ Cal}}{°C \text{ liter}}$$

Glass
$$\frac{0.20 \text{ Cal}}{°C \text{ liter}}$$

Steel
$$\frac{0.11 \text{ Cal}}{°C \text{ liter}}$$

How many Calories would it take to raise the temperature of 1 liter of steel from 15°C to 20°C?

$$\Delta T = 20°C - 15°C = 5°C$$

$$\frac{0.11 \text{ Cal}}{°C \text{ liter}} (5°C) = 0.55 \text{ Calories for 1 liter of steel}$$

What is this in joules?

$$(0.55 \text{ Calories})\left(4{,}148 \frac{J}{\text{Calorie}}\right) = 2{,}281 \text{ J}$$

This is a lot less than 20,740 J for 1 liter of water.

If you add heat to steel, its temperature will go up rapidly because it takes only 0.11 Calorie to raise steel 1°C. it takes almost 10 times more heat to raise the temperature of a liter of water 1°C. This makes water a great heat stabilizer. The earth is about $\frac{3}{4}$ covered with water. This helps move cold water toward the equator from the poles and warm water from the equator to the poles. It is sad when people, even Christians, dwell on negative aspects of life and do not see the great blessings God has bestowed upon everyone through His creation. Water in your body stabilizes your body temperature at right around 98°F. The heat generated by an automobile engine is dissipated by water in the cooling system. The structure and properties of a water molecule demonstrates God's wisdom and grace. Many give the glory to a false deity called mother nature rather than the true living God who paid the ultimate sacrifice for their salvation.

Matter undergoes phase changes where solids become liquids and liquids become gases. When ice is heated, heat melts ice into liquid water first and afterward additional heat raises its temperature. It takes approximately 0.08 Calorie to melt 1 gram of ice. This is called the latent heat of fusion (L_f) of ice. Ice becoming liquid water is a phase change. About 0.54 Calorie is required to cause 1 gram of water at 100°C to become steam. This is called the latent heat of vaporization (L_v). Other substances have different values for L_f and L_v.

Boiling point is defined as the temperature at which

Ice (solid)
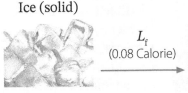

L_f
(0.08 Calorie)

Water (liquid)

L_v
(0.54 Calorie)

Steam (gas)

the vapor pressure (pressure of vaporizing molecules pushing against the pressure of the atmosphere) exceeds the atmospheric pressure. When you see recorded boiling points of different substances, they are assumed to be at sea level. As you go up a mountainside, the atmospheric pressure decreases because the layer of air above you becomes thinner. Therefore, your ears "pop." Your inner ear inside your eardrum is connected to the roof of your mouth by a tube called the Eustachian tube. The Eustachian tube has a mucus plug that prevents bacteria and other things from getting up into your inner ear. When you swallow, you suck the mucus plug out and air pressure in your inner ear equalizes with the outside air pressure. When the air pressure in your inner ear is greater than the outside air pressure, your eardrum bulges outward which distorts your hearing. When you swallow, the pressure is relieved, and your eardrum returns to its normal shape and you hear better.

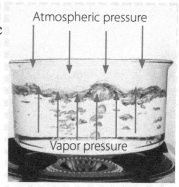

When you get to your cabin in the mountains, you set up camp and start a fire in your pot belly stove to cook your dinner. The first thing you do is boil some water. A curious thing happens. The water boils at less than 100°C. The air pressure is not as great up on the mountain, so the vapor pressure of the water in your pot does not have to be as great to exceed the atmospheric pressure. You also must cook your food in the boiling water longer because it takes longer to raise the temperature of the food before you can eat it.

As evening comes on and the sun goes down, the mountain air gets much colder than you have been used to. The fire in your pot belly stove heats the cabin nicely. You place a metal pan with spaghetti on top of the stove because you are still hungry. It must be the fresh mountain air. Heat is transferred from the hot metal surface of the stove to the bottom of the metal pan. This is conduction — the transfer of heat energy from the contact of a hotter object and a colder object. You place your hands above the stove and feel the hot air rising from the surface of the hot stove. This is convection — the rising of warm air carrying heat from a hot surface. As the air heats, the air molecules move faster as they gain more kinetic energy. As they move faster, they spread out into a greater volume. This lowers the density of the air above the stove — fewer air molecules in a cubic centimeter. The lighter warm air rises as the cooler heavier air sinks.

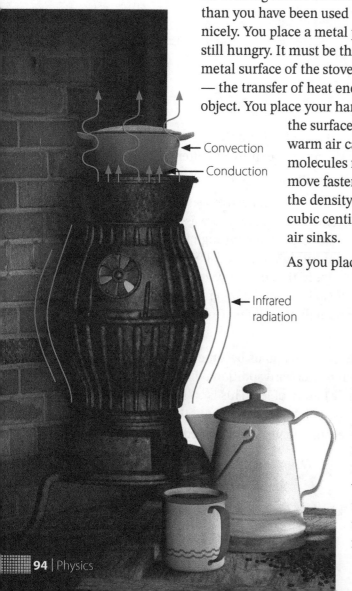

As you place your hands in front of the stove, you feel heat. This is not conduction, because you are not touching the stove unless you back into it. It is not convection, because warm air rises and does not go horizontally away from the front of the stove. This heat is called radiant heat. It is a form of light, but not within the range of visible light. It is infrared (IR) radiation. The light spectrum that we can see goes from red to orange to yellow to green to blue to violet. Red light has the least energy and violet has the greatest energy. Infrared means below red. It has less energy than red and is invisible to us. This is the wavelength of light that we feel as radiant heat. When an object has a higher temperature, it gives off more IR radiation. Satellites that measure the temperature of the ocean surface are measuring the amount of IR radiation from the ocean.

The three means of transferring heat energy are conduction, convection, and IR radiation.

If you live in a cold climate and have a greenhouse (a building with transparent walls and roof) to grow plants, ultraviolet (UV, greater than violet) radiation from the sun passes through the transparent walls of the greenhouse and heats everything inside. As the temperature within the greenhouse increases, objects in the greenhouse give off more IR radiation. The IR radiation cannot pass through the walls of the greenhouse, so the air and objects in the greenhouse get warmer. This is called the greenhouse effect.

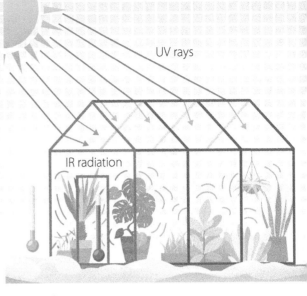

A person's view of origins has an interesting effect on how the greenhouse effect is interpreted in explaining weather patterns. Those who hold to a very old earth view global warming as a long-range phenomenon. They attribute increases in CO_2 and other greenhouse gases as trapping more heat within earth's atmosphere. However, if the earth is thousands rather than billions of years old from a literal interpretation of Genesis, we would still be in a warming phase from the effects of an ice age after the flood of Noah's day. If an old earth view is adopted, then there would not have been a literal worldwide flood because the sedimentation that formed most of the fossils would not have been formed by massive flooding because they would have formed over long ages. In this view, ice ages would have occurred too long ago to affect the climate today. With a young earth model, CO_2 and other gases (including O_2) would not have trapped IR radiation long enough to give substantial earth warming. In the history of earth, there have been numerous cycles of warming and cooling. An ice age would have to have had a massive buildup of moisture and shielding from the sun — such as could have occurred in the volcanic upheaval in the aftermath of the flood. When the effects that caused the ice age tapered off, the earth would have had gradual warming.

Slower moving water molecules in cold air are more likely to attach to dust and other particles in the air, forming water droplets that can become heavy enough to fall as rain, mist, fog, or snow. When air temperature drops, it cannot hold as much water and some of the water molecules settle out onto cold surfaces (where the air would be even colder) as dew. When water goes from a vapor to a liquid, the process is called **condensation**.

When the air has all the water vapor it can hold, it is saturated. When air cools at night it cannot hold as much water and some of it forms condensation. When the air warms up again as the sun comes up, it can hold more water and evaporation occurs.

The amount of water vapor is expresses as relative humidity. This is the amount of water vapor in the air divided by the maximum that air could hold at that temperature multiplied by 100 to give a percentage. If the relative humidity is 75%, the air has 75% of the water vapor that it can hold at that temperature. As the air temperature drops at night, the relative humidity will rise, even though no more water vapor has been added. This is because colder air can hold less water. When it gets cold enough for the water vapor to be more than the air can hold, some of it will condense on cool surfaces or on dust particles to form rain or fog droplets. Therefore, it rains more in the mountains where air is cooler at higher altitudes.

Temperature is caused by the velocities of atoms and molecules. Water is a special molecule because of its high heat capacity. It takes more joules to raise the temperature of water 1°C than most substances because the hydrogen bonding between water molecules impedes the motion of water molecules. This is why water is such a blessing in maintaining body temperatures.

LABORATORY 11

Heat Energy, Freezing Point of Water, and Relative Humidity

REQUIRED MATERIALS

- 3 containers: 1 with warm water, 1 with cold water and 1 with water at room temperature
- 1 cup size container with water
- 1 cup size container with water with as much salt as possible dissolved in it
- Freezer
- 2 thermometers
- Cotton cloth sleeve for the bulb end of one of the thermometers

Introduction
Our sensations of hot and cold are interpreted subjectively by our brains. Something feels cold because it has less heat energy than something else. These sensations are important to protect us from dangerous extremes. While conducting tests to determine the status of equipment or our state of health, more objective measurements are needed. A thermometer measures the expansion or contraction of mercury or alcohol with changes in heat energy. These measurements are repeatable and not subject to our feelings.

When something freezes, it undergoes a phase change. Temperature changes occur as heat energy enters or leaves something. When a phase change occurs, the temperature does not change. When water freezes, water molecules align themselves in a crystalline arrangement where there is much less movement of individual molecules. If something interferes with the alignment of the water molecules, more heat energy must be removed before they will form ice.

Relative humidity is the percent of saturation of air with water vapor. If the relative humidity is 30%, the air can absorb another 70% before it is saturated. In this procedure, the temperature of a wet bulb (the bulb of the thermometer is covered with a wet cotton sleeve) thermometer is compared to the temperature of a dry bulb thermometer. If the relative humidity is low, more water will evaporate from the wet bulb and lower its temperature. If the relative humidity is high, very little water will evaporate from the wet bulb and its temperature will not decrease as much.

Purpose
This laboratory procedure demonstrates the subjectivity of our senses of hot and cold, the effect of materials dissolved in water upon its freezing rate, and the relativity humidity as shown by the evaporation of water.

Procedure

 observe

1. Place one hand in a container with warm water and the other hand in a container with cold water. Hold your hands in the containers for at least 3 minutes. Take both hands out of the containers and place them both in a third container with water at room temperature.

 question

2. What does each hand feel like?

 research

3. Does the same water feel different to your separate hands?

 hypothesis/experiment

4. Our interpretation of the sensation of temperature is by comparisons, while thermometers detect temperature by the expansion and contraction of mercury or alcohol.

 analyze

5. Did this procedure confirm how you interpret sensations of temperature?

 conclusion

6. Explain.

 observe

7. Place water in two cup-sized containers. In one container dissolve as much salt

in the water as you can. Place both containers in the freezer and check them every 15 minutes to determine how long it takes the water in each container to freeze. Do not leave the freezer door open any longer than you must.

question

8. Which froze the quickest?

research/hypothesis

9. Design a hypothesis to explain their relative freezing times. A hypothesis is a possible explanation for your observations.

experiment

10. Design an experiment to test your hypothesis. Conduct the experiment and record your results.

analyze

11. Your experiment can either support your hypothesis or reject it.

conclusion

12. You cannot prove your hypothesis to be correct because there could always be one that has not been thought of.

observe

13. Construct a cotton cloth sleeve and slip it over the bulb of one of the thermometers. Tape the cotton sleeve in place. In the middle of the day, record the air temperature using the thermometer without the cotton sleeve. Soak the cotton sleeve on the other thermometer in water. Shake this thermometer back and forth in the air 25 times so that water evaporates from the cotton sleeve. Record the temperature of this thermometer. The evaporation of water from the cotton sleeve should have lowered the temperature of this thermometer as water molecules removed heat energy from the cotton sleeve as it evaporated. If the relative humidity is greater, less water will evaporate.

question

14. If the relative humidity is lower, will more water evaporate?

research

15. Use the following table to determine the relative humidity. Down the left side of the table are the air temperatures as measured by the dry thermometer. Across the top are the values from the dry bulb thermometer minus the temperature from the wet bulb thermometer.

hypothesis

16. Where these values intersect (as shown by the darkened row and column of the table) is the relative humidity. The example would be a dry bulb temperature of 20°C and the difference between the dry and wet bulb thermometers as 6 indicating a relative humidity of 51%. This means that the air at this time has 51% of the moisture that it could hold.

experiment

17. Determine the relative humidity just as the sun goes down and again first thing tomorrow morning.

analyze

18. Each time, record the time of day, weather conditions, and the relative humidity.

Dry-Bulb Temp., °C	Dry-Bulb Temperature Minus Wet-Bulb Temperature (Dry-Bulb Depression), °C													
	1	2	3	4	5	6	7	8	9	10	12	14	16	18	20
2	84	68	52	37	22	8									
4	85	71	57	43	29	16	3								
6	86	73	60	48	35	24	11								
8	87	75	63	51	40	29	19	8							
10	88	77	66	55	44	34	24	15	6						
12	89	78	68	58	48	39	29	21	12						
14	90	79	70	60	51	42	34	26	18	10					
16	90	81	71	63	54	46	38	30	23	15					
18	91	82	73	65	57	49	41	34	27	20	7				
20	91	82	74	66	59	51	44	37	31	24	12				
22	92	83	76	68	61	54	47	40	34	28	17	6			
24	92	84	77	69	62	56	49	43	37	31	20	10			
26	92	85	78	71	64	58	51	46	40	34	24	14	5		
28	93	85	78	72	65	59	53	48	42	37	27	18	9		
30	93	86	79	73	67	61	55	50	44	39	30	21	13	5	
32	93	86	80	74	68	62	57	51	46	41	32	24	16	9	
34	93	87	81	75	69	63	58	53	48	43	35	26	19	12	5
36	94	87	81	75	70	64	59	54	50	45	37	29	21	15	8
38	94	88	82	76	71	66	61	56	51	47	39	31	24	17	11
40	94	88	82	77	72	67	62	57	53	48	40	33	26	20	14
42	94	88	83	77	72	67	63	58	54	50	42	34	28	21	16
44	94	89	83	78	73	68	64	59	55	51	43	36	29	23	18

CHAPTER TWELVE

LAWS OF THERMODYNAMICS

OBJECTIVES

At the conclusion of this lesson the student should have an understanding of

- System and surroundings
- Changing internal energy by heat and work
- The First Law of Thermodynamics
- Carnot Cycle
- Efficiency of a heat engine
- The Second Law of Thermodynamics
- Entropy
- The Third Law of Thermodynamics

Thermodynamics means heat movement. You have quite a bit of background in this area by practical experience. Sometimes it is hard to recognize the obvious by the vocabulary. That is certainly the case for this area of physics. As you study this lesson, the worksheet will serve as a review and study guide. Read through this chapter slowly and think through what you are reading. Afterward, the questions in the worksheets in the teacher guide will guide you through what you need to know for the quiz and later exam. For each of the laws of thermodynamics, think about it in terms of your own experience. They were developed in order and each contributed to the next.

In the 1800s, the question came up — can an object have an increase in temperature without touching a hotter object? Benjamin Thompson, named Count Rumford by the King of Bavaria, was involved in weapons development. He noticed in 1798 that as they bore out the opening in a cannon, the drill bit and the cannon became very hot. The temperatures of the cannon and drill bit were increasing without touching a hotter object. Somehow the work being done on the cannon was being converted into heat energy. He proposed a hypothesis that mechanical work could in a consistent and predictable way raise the temperature of a system (the cannon and drill bit).

Count Rumford and one of his original sketches

In these discussions, the word **system** refers to what is being considered and the word **surroundings** refers to what is around the system. For example, the cannon with the drill bit was the system and the surroundings absorbed some of the heat produced.

Later in the 1840s, James Prescott Joule devised a simple but clever experiment to test the hypothesis. He had a paddle wheel in water connected by pulleys to a hanging weight that fell and turned the paddle wheel. The paddles moved the water molecules, which increased their kinetic energy, which raised the temperature of the water. A thermometer was placed in the water to measure the temperature change (*Diagram 12.1*).

He measured the energy turning the paddle wheel by the difference in the gravitational potential energy of the falling weight. If the mass of the weight was 1 kg, and it was at an initial height of 1.1 meter and fell to 0.1 meter, the initial potential energy would be . . .

$$mgh = (1 \text{ kg})\left(9.8 \tfrac{m}{s^2}\right)(1.1 \text{ m}) = 10.78 \text{ J}$$

and its final potential energy would be . . .

$$mgh = (1 \text{ kg})\left(9.8 \tfrac{m}{s^2}\right)(0.1 \text{ m}) = 0.98 \text{ J}$$

and the change in potential energy would be 10.78 J – 0.98 J = 9.8 J.

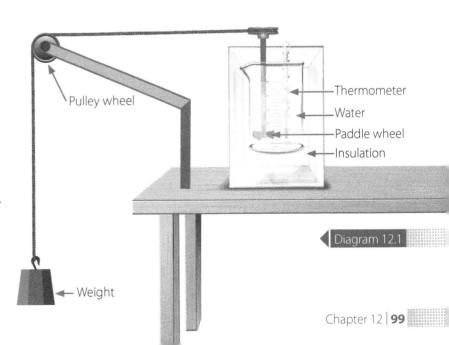

Diagram 12.1

That would give a total possible energy available to raise the temperature of the water to be 9.8 J or 9.8 J $\left(\frac{1 \text{ Calorie}}{4{,}148 \text{ J}}\right)$ = 0.0024 Calorie or 2.4 calories. If the container had 1 liter of water, it could have raised the temperature of the water by 0.0024°C because 1 Calorie will raise 1 liter of water 1°C. This is not much of a temperature rise, but consider that it came from the potential energy difference of only 1 meter. If a weight of more mass was used and allowed to fall a greater distance, the rise in temperature would be greater. The hypothesis made earlier by Count Rumford was supported.

Energy produces work and it was demonstrated that work as mechanical energy could be converted into thermal energy. The connection is that the temperature rise was an increase in the motion of the water molecules that was caused by the motion of the paddles.

This and other experiments led to the idea that energy and work could be measured and accounted for. This led to the development of the First Law of Thermodynamics by William Thomson (Lord Kelvin) in the late 1800s that energy was conserved — it could not be created or destroyed (except by God). This meant that dependable measurements could be made. Energy would not just disappear or appear out of nowhere. These things we take for granted today, but they were just being discovered back then. This is another example of the order that God planned into this universe. When something lost some energy (its temperature dropped) something else gained the same amount of energy.

It has been shown that there is a relationship between matter and energy where matter can become energy. This was the basis of Albert Einstein's equation . . .

$$E = mc^2$$

where m is the mass of matter and c is the speed of light and E is the amount of energy released. This became the basis for the energy release in the atomic bomb.

Another way that it was expressed was . . .

The increase in the internal energy of a system is equal to the amount of heat added to a system minus the amount of work done by the system on the surroundings.

The First Law of Thermodynamics is also called the Law of Conservation of Energy.

It would be like saying that the change in the amount of money in your checking account equals what you put in minus what you took out.

This is usually expressed by the expression . . .

$$\Delta U = Q - W$$

ΔU is the change in the internal energy of a system, Q is the amount of heat added to the system, and W is the work done by the system on the surroundings. This was important for the applications to heat engines. If you added heat (Q) to the system, it would increase the internal energy (U) of the system. This happens when you heat water to raise its temperature. When Joule had his paddle wheel in water, he increased the internal energy (U) of water by doing work ($-W$) on it. The work is negative because it is being done **on** the system, not by the system on the surroundings. If you put $-W$ into the above equation, you get $-(-W)$ which is $+W$, where $\Delta U = Q - (-W) = Q + W$. work done on the system raised the internal energy of the system.

Prior to this, Robert Street, in 1794, patented an internal combustion engine that used petroleum. The idea that you could measure and control energy was important for building more efficient engines.

In 1824, the French engineer Carnot published a paper on what an ideal gas engine would be like. This was not a real engine but a model of what could happen if the heat added, minus the work done, minus the heat lost as exhaust, would equal the starting temperature. If it worked, it would be reversible and could go in either direction. It turned out that it could not be 100% efficient because some heat was always lost to the exhaust that could not be reclaimed. This was used by Kelvin in 1850 to come up with the second law of thermodynamics.

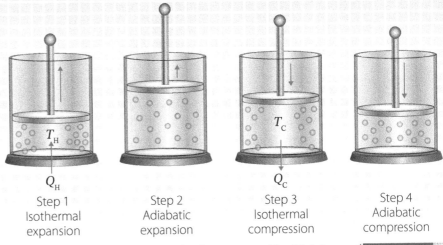

Diagram 12.2

Carnot's model is called the Carnot Cycle as shown in Diagram 12.2.

In step 1, the temperature of the gas in the cylinder is raised by adding heat to a high temperature T_H. The gas expands, pushing up on the piston. While the gas expands it remains at the same temperature (isothermal). This is the step resulting from the surroundings adding energy (Q) to the system. Isothermal means at the same temperature.

In step 2, called adiabatic expansion, the gas continues to expand with no new energy input from the surroundings. Adiabatic means no heat flow.

In step 3, work is done by the surroundings (−W) on the system (gas in the cylinder) compressing the gas. This is called the exhaust step because it returns the gas to its original temperature by getting rid of the extra heat (−Q). This is called isothermal compression.

In step 4, because the temperature of the gas is lower, the piston can compress it further. This is called adiabatic compression because no heat is flowing between the system (gas) and the surroundings.

U = internal energy
Q = heat
W = time
T_H = high temperature
T_C = low temperature
e = efficiency

After the first law of thermodynamics was developed, the efficiency of the Carnot Cycle could be evaluated. After calculating the work done by the engine in steps 1 and 2 and the work done by the surroundings on the gas in steps 3 and 4, the amount of heat added in step 1 and removed in step 3, the efficiency was calculated to be . . .

$$e = \frac{W}{Q_H} = \frac{T_H - T_C}{T_H}$$

This would be the maximum efficiency that an ideal engine could have if it had a high temperature T_H in step 1 and a low temperature T_C in step 3.

The temperature scale was in Kelvins, developed by William Thomson (Lord Kelvin) in the 1850s. The Kelvin scale is °C + 273.15.

If the gas in the cylinder took in heat in step 1 at 327°C and gave up heat in step 3 at 27°C, T_H would be 327°C + 273 = 600K and T_C would be 27°C + 273 = 300K.

$$e = \frac{600K - 300K}{600K} = 0.5 \text{ or } 50\% \text{ efficiency}$$

This means that an ideal engine operating between these temperatures could have a maximum efficiency of 50%. This means that only 50% of the heat energy put into this engine could be converted into work. The energy that is converted into work is called free energy.

What if T_H were 700K instead of 600K?

$$e = \frac{700 - 300}{700} = 0.57 \text{ or } 57\% \text{ efficient.}$$

What if T_H were 800K instead of 600K?

$$e = \frac{800 - 300}{800} = 0.625 \text{ or } 62.5\% \text{ efficient.}$$

The higher you raise T_H, the closer you would come to 100% efficient, but you would never reach it. This meant that when the temperature of the gas in the cylinder was raised to T_H in step 1, not all the thermal energy could be converted into work in steps 1 and 2. Some of the heat had to leave as exhaust. This meant that he did not have a perpetual motion machine that would keep operating with no added heat. He had to keep putting additional heat energy into the gas in the cylinder in step 1 to keep the engine running. This was summarized in the **Second Law of Thermodynamics** where every time energy is transferred to or from the system, some energy is lost to the surroundings. This is why automobile exhaust is so hot.

Up to now, you may have been wondering why we went through the confusing Carnot Cycle. The reason is to show where the understanding of the efficiency of an internal combustion engine came from. Since Henry Ford put the first Model A automobile into production in 1903, automotive engineers have been trying to find various ways to improve the efficiency (mileage) of automobiles and reduce environmental pollution from the exhaust.

Using Carnot's model of an ideal engine, Kelvin came up with the **Second Law of Thermodynamics**. He said that . . .

> No engine, working in a continuous cycle, can take heat from a reservoir at a single temperature and convert that heat completely to work.

Another way to state it is to say that heat always goes from a hotter object to a colder object and never from a colder object to a hotter object. The colder object becomes warmer — its thermal energy increases. In so doing, its state of disorder increases. The word **disorder** refers to the probability of finding something. When the atoms and molecules move faster because they gain thermal energy, it is harder to know where to find them. Let us say that there are 5 cats sleeping in the tall grass of a vacant lot. Suddenly there is a loud sound from a nearby construction project — the cats wake up and start running all over in the grass. When they were asleep, you knew where to find them. Now it is more difficult to find them — especially because they are constantly moving. We can say that the cats are more disordered because it is harder to find them. When you heat a gas, the molecules have greater kinetic energy and, if they can, they move out into greater volume. The state of disorder of the molecules has increased because it is more difficult to find them.

If you open a bottle of strong perfume, the molecules will diffuse out into the room as evidenced by the strong odor. The state of disorder of the perfume molecules has increased greatly. Try putting them back into the bottle! If you could, it would take considerable effort and energy.

In the Carnot Cycle, to start again at step 1 after step 4, additional energy would have to be supplied to make up for the lost exhaust.

The measure of disorder is called **entropy**. The letter S is used for entropy where . . .

$$\Delta S = \frac{\Delta Q}{T}$$

The greater the energy transfer (ΔQ), the more disorder, entropy, results. In the entropy equation, T is in the denominator because, as it gets greater, the less the entropy (disorder) will be because the change in kinetic energy will be less

The first law of thermodynamics has been called a law of nature. The second law of thermodynamics has been called a statement of probability. The greater the disorder, the less the probability of going back to the original order.

Low entropy
Low disorder

High entropy
High disorder

This has been applied to evolutionary thought in that the probability of smaller molecules assembling into larger molecules with specific order is very unlikely, with very small odds. The second law of thermodynamics is against order, (well-ordered) large proteins, and other molecules of living cells) coming from disorder. The claim is made that even if the probabilities are against the evolution of molecules of living cells, the worst probability can happen given enough time. Therefore, an old earth is so important to evolutionary ideas. But as the state of disorder increases, it will take apart everything that was formed. To restore order (return the gas in the cylinder to T_H) energy input from the surroundings is necessary. The sun is cited as the source for that energy for the evolution of large molecules. But, in addition to energy there must be a means to control the energy and put the smaller molecules together in specific arrangements. The probability of them coming together in the right order would not happen often enough to supply enough good molecules. Also, the energy would have to be controlled so that it would not destroy any molecules that might have formed.

It is said that entropy increases everywhere except in living cells. This, however, does not mean they could have evolved from less-ordered molecules. Living cells were created with enzymes, coded for by DNA, that physically order the assembly of molecules within cells. The energy from metabolism is captured and released in small manageable amounts by molecules called ATP (adenosine triphosphate) within organelles called mitochondria within cells. Cells began with a great degree of order in their cellular and molecular structures. When a system has an increase in order, as occurs within our cells, the surroundings have an increase in disorder. Therefore, we give off heat energy and waste, which increase the entropy of our surroundings. The second law of thermodynamics is still a strong argument for creation. You may run into these arguments in future studies as scientific evidences against creation. This study will prepare you to deal wisely in discussing your faith as Paul did in Athens.

As introduced earlier, Lord Kelvin developed the Kelvin scale in the 1850s. This came about by the study of gas pressures and temperatures. It can be shown by the pattern that as the temperatures of gases are reduced, their pressures decrease. **Pressure** is the force the gas molecules exert on the sides of their container, which is force per unit area. It can be expressed as pounds per square inch or Newtons per square meter. The pressure of a gas on the walls of its container decreases as the temperature drops because the average kinetic energy of the molecules hitting the sides of the container decreases. A plot of temperatures versus pressures for different gases is shown in this diagram (*Diagram 12.3*).

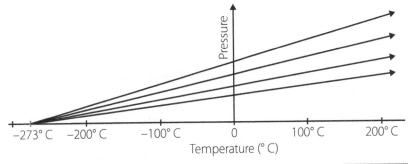

Diagram 12.3

Lines connecting temperatures of each gas extended to the left come together at zero pressure at a temperature of –273.15°C. This is where the value of **absolute zero** came from. The **Third Law of Thermodynamics** states that matter cannot attain absolute zero. At this point, the gas pressure would be zero, meaning that it has no thermal energy. All motion within the gas would cease. This has never been attained (even though some have come very close), so it is a consistent observation.

Thermodynamics is the study of the movement of heat energy. The transfer of heat in heating and cooling and its applications are better understood by the three laws of thermodynamics.

Heat energy increases the kinetic energies of particles causing them to do more work. This is the basis of internal combustion engines of we are greatly dependent. Heat energy is supplied by the Sun which is directed and controlled by systems created by God – such as photosynthesis whereby plants capture energy from sunlight to convert carbon dioxide and water into sugar and oxygen.

Laboratory 12

Second Law Of Thermodynamics

REQUIRED MATERIALS
- 4 foam cups (specifically polystyrene foam)
- Thermometer
- 2 100 ml graduated cylinders
- Source of hot and cold water

Heating elements raising temperature and melting ice on glass

Introduction
Heat only goes from a hotter object to a colder object. When heat is transferred, some of it is lost to the surroundings.

Systems go to a state of greater disorder where molecules and energy diffuse to all the region available. Greater disorder means greater difficulty in finding something as when gas molecules diffuse out into a greater space.

If you take 5 nickels and add 5 more nickels, you get 10 nickels. When you add Calories from hotter water to the Calories of colder water, you get less than the total. Some of the Calories have gone from the system (water samples) to the surroundings, increasing the disorder of the surrounding air molecules. The kinetic energy of the air molecules is increased, causing them to diffuse farther out.

In this exercise, you will use a graduated cylinder (tall tube with volume markings) to measure 100 ml (milliliters). The measured hot and cold water are poured into the foam cups.

Purpose
This exercise demonstrates the measurable loss of heat energy when heat is transferred from hotter water to colder water as described by the second law of thermodynamics.

Procedure

 observe

1. Place 1 foam cup into another foam cup. With the graduated cylinder, add 100 ml of hot (not boiling) water to the cup. Take the temperature of the hot water in degrees Celsius.

 question

2. Repeat step 1 with 2 more cups. This time add cold water to the cup and take its temperature in degrees Celsius.

3. Pour the contents of both cups into a bowl and take the temperature of the mixture in degrees Celsius.

 research

4. Calculate the number of Calories in the original hot water.

    _____°C × 0.10 liter × $\frac{1 \text{ Calorie}}{\text{°C liter}}$ = _____ Calories

5. Calculate the number of Calories in the original cold water.

    _____°C × 0.10 liter × $\frac{1 \text{ Calorie}}{\text{°C liter}}$ = _____ Calories

6. Calculate the number of Calories in the mixed hot and cold water.

    _____°C × 0.20 liter × $\frac{1 \text{ Calorie}}{\text{°C liter}}$ = _____ Calories

7. Add the Calories for the original hot and cold water samples.

    _____ Calories (step 4) + _____ Calories (step 5) = _____ Calories

 hypothesis/experiment

8. How did the number of Calories in the mixture compare to the sum of the hot and cold water samples?

 analyze

9. How would you explain your results from the second law of thermodynamics?

 conclusion

10. If you lost some Calories, where did they go?

CHAPTER THIRTEEN
WAVES

OBJECTIVES

At the conclusion of this lesson the student should have an understanding of

- Waves as energy
- Circles in water waves
- Wave trains
- Standing waves, simple harmonic waves, transverse waves, longitudinal waves
- Parts of a wave
- Wave interference and diffraction

A **wave** is energy moving from one place to another. If you took 100 dominoes and stacked them on end in a line so that they are close to each other, then push over the first domino, it will hit the second domino knocking it over, which causes that one to hit the next one, and they keep falling over until the last one falls over. All you did was to push over the first domino. The rest fell over down the line. Two things were present. There was a force field — gravity. Second, each domino had a potential energy because it was on its end and could fall over if pushed.

With ocean waves, water molecules are pushed by wind, which causes them to pile up and fall forward. They hit those in front of them causing them to fall forward. This pattern repeats itself until there are no longer any water molecules to fall into (the other side of the ocean). This is like the dominoes that keep falling over until the last one falls over.

When water molecules are pushed forward, they go down and are pushed back up by the ones on top going forward. This is because water molecules are small and move past each other — quite unlike dominoes. This causes water molecules to go in circles — forward, down, back, up, and forward again. Have you noticed by the shoreline, when the surf breaks it looks like part of a circle coming around and falling over (*Diagram 13.1*)?

In both dominoes and water waves, they are waves of energy. This is also seen in light waves and sound waves.

This may sound confusing compared to the usual image of a wave. Ocean waves have a **wave train** that looks like the up and down motion that we are used to seeing. This is formed by the positions of water molecules in successive circles. Water molecules are in different positions of their circles as they form a wave train. This gives the appearance of the up and down pattern of a wave.

◁ Diagram 13.1

Water molecules in their position in their circle push the water molecules in the next circle, which push the water molecules in the next circle. Water waves are also called gravity waves, because gravity provides the force to move the water molecules once they are pushed by the ones behind them, just like the dominoes. The effect is that the water molecules keep going around their circles. An ocean wave train can travel all the way across the Pacific Ocean, losing little if any energy. A storm in Asia can start a wave that crashes up on the shore of Southern California. A strong force, like an

Chapter 13 | 107

underwater earthquake in Asia, can cause huge circles of water that are formed across the ocean. Therefore, there are signs along west coast beaches warning of possible **tsunamis** (huge devastating waves).

Tsunami waves form huge circles in deeper water. As the wave train approaches shallower water near shore, the huge circle starts to flatten out because the sea floor changes its shape. The circle becomes taller and narrower until it is very tall and dangerous when it reaches the shoreline. Some of these have been very devastating. When there is an underwater earthquake, there is an automatic warning system that alerts those on different shores around the ocean. This is because the wave can travel all the way across the ocean with very little energy loss.

The circles on the surface cause circles to form below them. Each circle gets flatter the deeper they go, until the deeper ones are water molecules going back and forth.

Besides a tsunami getting taller as it approaches shore, regular waves do the same thing. The circles of waves get flatter on the bottom and taller of the top as the shallower sea floor alters the wave circle. When large circles come on shore, they have higher crests, to a surfer's delight. This is where they mount the waves and ride them into the shore. As they get closer to shore, the circles get taller and then suddenly, when the shallow bottom takes the bottom out of the circle, the surf rider comes over the top into the foam of the surf zone.

God's creative design is more than technical detail, which is great, but it is also for enjoyment and beauty. Look around at God's creation. Don't miss the grandeur of it. I remember the first time I saw Lake Michigan — it is huge! If a cursed world can be so beautiful and function with precision, can you imagine what the new heavens and earth will be like? Take time to enjoy the water and to watch a sunset over the water. You can enjoy a work of beautiful art so much more when you know the Artist. If you take the time and ask Him, He would be delighted to show you the great things He has created — especially the simple things in everyday life.

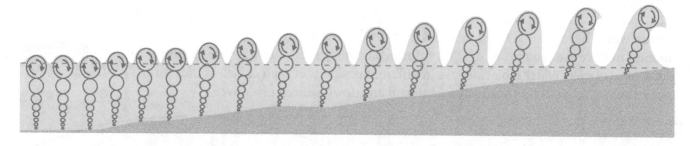

Radio waves are electromagnetic waves that can travel through space. They can even travel around objects in space. So it's interesting that the *Voyageur* spacecraft that are in interstellar space beyond our solar system are still sending back radio waves that reach us across the great expanse of space. Their signals are not losing energy, but they are spreading out by the Inverse Square Law that we looked at in an earlier chapter.

If you tie a rope securely at one end and move the other end up and down, you can produce a wave pattern in the rope. If you move the end of the rope up and down, you produce a pattern that looks like this.

It will look like the rope is wide in some areas and standing still at points in between. This is called a **standing wave** because it looks like the rope is standing still. It is just moving fast. Even here, the material of the rope is not moving away from your hand, but energy is. With solid materials like guitar string, piano wires and ropes, wave interference is caused by the wave going from your hand to where it is attached, bouncing back over the rope causing constructive interference in the middle of the rope. This is the interference of waves caused by going in opposite directions. Where the rope appears not to be moving is called the **node** and where the wave appears to be wide (due to constructive interference) is the **antinode** (*Diagram 13.2*).

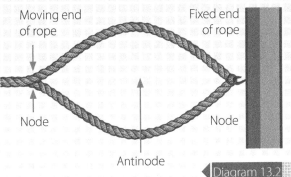

Diagram 13.2

In trigonometry, the shape of this wave is called a sine wave. Another name for it is a **harmonic**. If there are several identical up and down areas of the rope, it is called a **simple harmonic wave (SHW)**. The word *simple* means repeated in an identical manner. A wave pattern that goes up and down, like the rope, is called a **transverse wave**. A wave that goes forward and backward is a **longitudinal wave**, like sound waves. Electromagnetic waves (light) are both transverse and longitudinal.

The basic parts of a wave pattern are . . .

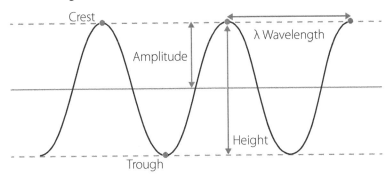

The top of a wave is the **crest** and the bottom of a wave is the **trough**. The distance between the middle of the wave and the crest is the **amplitude**. The distance between the crest and the trough is the height of the wave. The distance from one crest to the next is the **wavelength** symbolized by the Greek letter lambda λ.

The **frequency** (*f*) of a wave is the number of crests that pass a given point each second. It is expressed as $\frac{\text{cycles}}{\text{second}}$ or Hertz (Hz). It is given in units of $\frac{1}{\text{sec}}$ because cycles are not measurable units. Waves with greater frequency and shorter wavelength, have more energy.

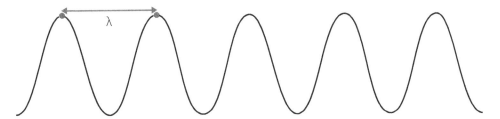

- Shorter wavelength (λ)
- Higher frequency
- More energy

Waves with lower frequency and longer wavelengths have less energy.

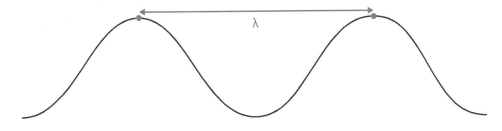

- Longer wavelength (λ)
- Smaller frequency
- Less energy

The units of frequency are $\frac{cycles}{second}$ and the units of wavelength are $\frac{meters}{cycle}$. Frequency times wavelength give the velocity of a wave. Remember, this is the velocity of the wave train, not the individual molecules. For example, the velocity that the wave travels down the rope, even though the molecules of the rope do not travel down the rope, is . . .

$$f \times l = v$$
$$\frac{cycles}{second} \times \frac{meters}{cycle} = \frac{meters}{second}$$

If two cars try to occupy the same space at the same time while traveling down the road, there will be problems (even though some do not seem to understand that). Because waves are energy, two or more waves can occupy the same space at the same time. Notice in the following diagram what happens when waves reflect from a curved shoreline. The lines in the diagram are the crests of the waves. They reflect back onto each other so that they cross over.

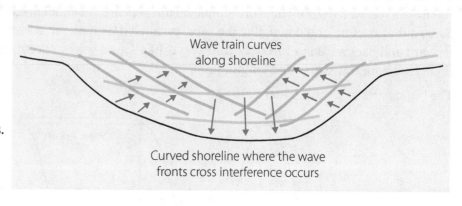

When waves overlap, a phenomenon called interference occurs. If the crest of a wave overlaps the crest of another wave, they add to each other and produce a crest that has the sum of the amplitude of both. This is also called **constructive interference** because the troughs add to each other. When the crest of a wave meets the trough of another wave, **destructive interference** occurs because they subtract from each other."

These can produce what are called **rogue waves**, where ships go way up on a high crest and then go rapidly down into a very deep trough. These have sunk some ships.

Crisscrossing wave patterns cause an X pattern in the water as the waves interfere with each other on Lake Michigan.

Waves can undergo **diffraction** where they pass through a narrow opening and fan out on the other side. This happens when waves hit a narrow opening in a breakwater that protects the shoreline from rough waves.

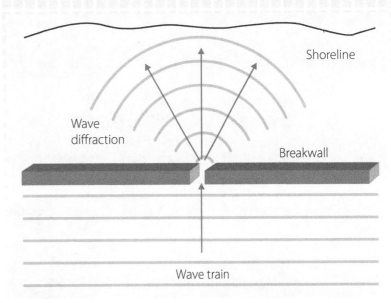

When this happens to light waves, diffraction breaks light down into its separate colors. This is because different colors have different energies that cause some to fan out more than others.

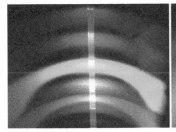 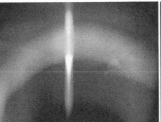  

Many of the following chapters dealing with sound, light (electromagnetic radiation), electricity, and magnetism are applications of these properties of waves. There are even wave behaviors in gravity force fields.

Above is the spectra of various light sources as revealed in a reflection diffraction-grating shown from left to right: fluorescent lamp, incandescent bulb, candle flame, and white LED lighting.

Chapter 13 | 111

LABORATORY 13

Water Waves

REQUIRED MATERIALS

- Tub with water
- 3 small rulers
- **Note:** You will also need an assistant for this lab.

Introduction

Water molecules move in a circular motion, giving the illusion that it is a wave pattern. This was first noticed by watching floating objects as a wave passed by.

The pattern of a wave moving across the surface of water is called a wave train to distinguish it from the actual movement of water molecules. Often the word *wave* is used, meaning the wave train.

If the crest of a wave meets the crest of another wave, the **wave height** is the sum of both crests. This is called constructive interference. If the trough of a wave meets the trough of another wave, the wave depth of the resulting trough is the sum of both troughs. This is also called constructive interference. If the crest of a wave meets the trough of another wave, the resulting wave height is the depth of the trough subtracted from the height of the crest. If the crest is just as high as the trough is deep, the water will appear to be flat at that place. This is called destructive interference.

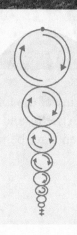

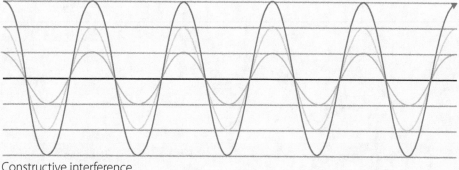

Constructive interference

Constructive interference

Destructive interference

Purpose

In this exercise, the student will observe the circular pattern of water waves and how constructive and destructive interference occurs in lakes and oceans.

Procedure

 observe

These procedures are done in a bathtub or similar body of water.

1. Place a ruler (or similar sized object) halfway submerged lengthwise in water. Move the ruler back and forth, producing a series of waves. Do it with the ruler parallel to the other side of the tub so that the waves hit the side and bounce straight back. Watch as the reflected waves interfere with the original waves.

 question

2. Can you see patterns of constructive interference and destructive interference?

 research

3. Write out a detailed description of what you observe, using complete sentences.

 hypothesis

4. How are you identifying areas of constructive and destructive interference?

5. Make a series of waves that hit the side of the tub at a 45° angle (approximate).

 experiment

6. What kind of pattern do you observe now? Describe it in detail. How were the patterns different from those in step 1?

7. Take 2 rulers and make waves from two sides going toward each other. Describe the patterns formed this time.

8. With 1 ruler, make waves going lengthwise the longest distance of the tub. Do the waves appear to lose energy as they go across the tub? Describe the pattern as the waves bounce off the far side of the tub and encounter the original waves that you are making.

9. Place a small floating object in the middle of the tub. Make waves from the farthest side of the tub. Describe the movement of the floating object as the wave passes it. Does it go all the way with the wave, or can you see circular motion in the water by the movement of the floating object?

10. Have someone help you by holding 2 rulers forming a breakwater with a small gap between the rulers.

    Make waves with a third ruler so that the waves hit the breakwater parallel to the breakwater. Describe the pattern of the waves that pass through the opening between the rulers and come out on the other end. Do you see a wave diffraction pattern? Explain.

11. Drop a marble or similar object in the water. Observe and describe the circular wave train that proceeds out from where the object was dropped. Drop two objects into the water a fair distance from each other and observe their wave patterns and how they intersect with each other.

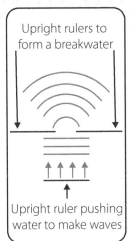

 analyze

12. Look carefully for areas of constructive and destructive interference.

 conclusion

13. This step is optional depending upon availability. Go to the shoreline of an ocean or lake and observe the wave trains as they approach shore and what happens to them when they reach shore. As the circles of water approach shore, the water gets shallower. The bottom sediment flattens the circular motion of the water, causing the bottoms of the circles to flatten out and the tops to becomes taller. Describe what you observe as the wave trains come on shore.

CHAPTER FOURTEEN

SOUND WAVES

OBJECTIVES

At the conclusion of this lesson the student should have an understanding of

- The nature of sound waves
- Frequency ranges of sound
- The refraction of sound waves
- Velocity of sound in different media
- Reverberation, forced vibrations
- Resonance
- The Doppler Effect
- Frame of reference

Sound was created at the beginning of creation. "God created the heavens and the earth" (Genesis 1:1). Sound is such a marvelous gift from God that we take for granted or just feel that it should be there, so we do not think of thanking God for it. This, however, is not the case, especially if you are deaf or have lost your hearing later in life. It is a tragedy that we do not appreciate until we lose it. It is such a blessing to hear birds first thing in the morning or the sound of waves lapping up on the shore.

Science — physics — is a gift from God that enables us to better understand and use the blessing of sound that He gave us. The ability to talk by phone, Skype, or FaceTime and to see and hear a loved one across the world at a moment's notice is great. We just must be careful not to let physics and the pleasures that it provides us take the place of the One who provided sound in the first place.

Sound waves are longitudinal waves that travel through a medium — usually air. Light waves can go through a vacuum, but sound waves cannot. If that were not the case, the universe might be a very noisy place. Even this demonstrates God's wisdom. The longitudinal waves of sound are like those produced in a Slinky™ when you push and pull back on it. You can see areas moving through the Slinky™ where it is compressed and where it is spread out.

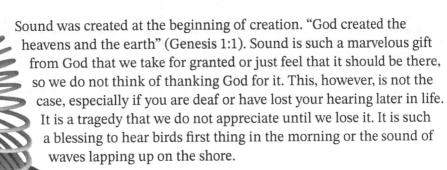

λ

When you speak, you compress the air in front of your mouth. This produces a pocket of air where the air molecules are closer together. As they spread out, they push on the air molecules in front of them, which push on the air molecules in front of them. These areas of compressed air, called condensations, keep pushing on air molecules making up the crests of a sound wave. When the air is compressed, it leaves behind a partial vacuum called a **rarefaction**. As the sound wave travels, it compresses the air next to your eardrum which vibrates. Your vibrating ear drum causes tiny bones (which act as amplifiers) to vibrate, which eventually cause nerve impulses to go to your brain. The sound you hear is your brain's interpretation of the nerve impulses. Can you imagine if the same air that left a person's mouth reached your eardrum? If you were in a room with many people talking at once, you would smell the bad breath of everyone talking. Yuck! The pulses of air leaving your mouth are caused by vibrations of your vocal cords in your pharynx at the top of your throat. Your vocal cords vibrate, causing the pulses of compressed air. This is not the same as blowing air from your mouth.

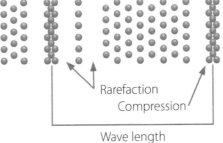

Rarefaction
Compression
Air molecules
Wave length

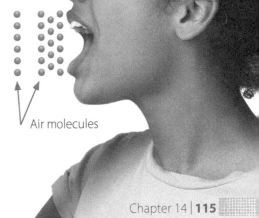

Chapter 14 | 115

Sound waves have the same properties as other waves as discussed in chapter 13. The areas of condensation are the crests of the waves and the areas of rarefaction are the troughs. Constructive and destructive interference occurs with sound waves coming from different sources or reflected from hard surfaces. The frequency of sound waves is the number of areas of condensation that pass a given point each second. We usually hear sounds from 16 Hz (hertz or $\frac{cycles}{second}$) up through 20,000 Hz. Higher frequency sounds have higher pitch (soprano) and lower frequency sounds have lower pitch (bass). Sound with frequencies below 16 Hz are called **infrasonic** and sound above 20,000 Hz are **ultrasonic**. Most music is below 4,000 Hz. When an ultrasound images are formed for medical diagnoses, sound waves greater than 20,000 Hz are used which can give soft tissue images that cannot be seen with x-rays. It is often used to check on the health of a fetus and to tell if the baby is male or female.

The velocity of sound traveling through air depends upon the temperature of the air. In warm air, the molecules are already moving faster, so the velocity of sound is greater. Likewise, sound travels more slowly in colder air. Therefore, when the air near the ground is colder and the air above is warmer, sounds are heard better near the ground. This is because as sound moves more slowly along the ground, the sound waves are bent toward the ground. In the diagram below, areas of condensation are shown by lines (like a wave train). As the sound waves are slower near the ground, they do not go as far in the same time as in the warmer air. This bends the sound waves toward the ground. This is an example of **refraction**, which is the bending of waves.

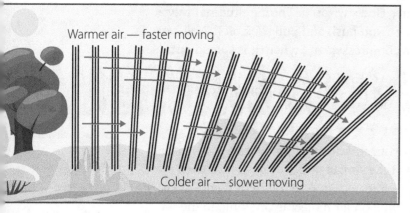

At room temperature, sound travels an average of 340 $\frac{m}{s}$ (750 $\frac{miles}{hour}$). A sound wave takes about 3 seconds to go a kilometer (5 seconds to go a mile). For every 10° C temperature rise, the velocity of sound increases by 6 $\frac{m}{s}$. Light is much faster than sound, so if you see a flash of lightning and hear the thunder 5 seconds later, the lightning is 1 mile away. If you heard thunder 10 seconds after seeing the flash, how far away is the lightning?

In water, sound travels about 4 times faster because water molecules are much closer to each other and they push on each other more readily. When you are at a swimming pool, have someone say something underwater and listen underwater. You will hear it much more readily than above water.

In World War II, German submarines were sinking American supply ships crossing the Atlantic. To find the submarines to destroy them, triangulation was used. This is where 3 microphones were placed n the water to listen for the sound from the engines of the submarines. By placing the microphones at 3 positions around the Atlantic Ocean, they could tell how far from each microphone a submarine was at and pinpoint its position.

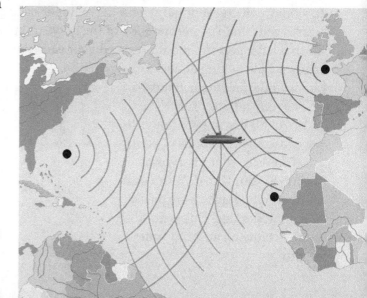

It had been discovered earlier that whales would call to each other over great distances in the Atlantic in a dense layer of water called the SOFAR channel. The sound would stay in that layer of water and spread out across the Atlantic. The microphones were placed in the SOFAR channels. It is amazing how God's gift to the whales helped save many lives in the war.

At about 700 meters deep water density keeps this layer from mixing and traps sound. SOFAR stands for Sound Fixing And Ranging

Sound travels much faster in solids, especially metals, because of their greater density. Have someone hit a long metal object with a hammer while you place your ear near the other end and listen for the banging sound. See if you can hear 2 bangs — one from the metal object that reaches your ear first and the second bang carried more slowly through the air.

A sound that is reflected from a hard surface that you hear again as it comes back to you is an **echo**. When there are many sounds at the same time in a room with hard surfaces, many reflected echoes can be heard called **reverberations**. This is not an ideal room for music performances or public speaking. Sometimes when sound systems or reverberations cause sound waves to overlap so that the condensations of one form destructive interference with other sound waves, they can produce dead zones where you cannot hear the sound. When that happens, there are usually other areas of constructive interference producing very loud sounds. This can be equally disturbing. This is where great care needs to be used in designing sound systems for auditoriums.

When two sound waves of slightly different frequencies are emitted together, repeated intervals of loud and faint are produced from constructive and destructive interference. The pulses of loud sounds are called **beats**. This is a very familiar term in music. When the beats are in the bass (low frequency range), they give the "boom boom" sound, especially when that is all you hear — like when a neighbor has music turned up too loud.

Forced vibrations occur when sound waves from the vibrations of one object cause vibrations in another nearby object. When you strike a tuning fork, its sound waves can cause vibrations in another tuning fork causing it to give off sound waves as well. A special case is where the **natural frequency** (a frequency at which an object naturally vibrates) of an object is the same as the frequency as the sound given off by another object. This is called **resonance**. If 2 tuning forks have the same natural frequency, sound waves given off by one of them will cause the other to vibrate at that frequency. In November 1940, the Tacoma Narrows Bridge south of Seattle, Washington, began to sway because the frequency of the pulses of wind matched the natural frequency of the bridge (a property of the material and thicknesses of the bridge). The wind caused the bridge to sway farther until it broke apart and fell into the water. This was a tragic example of resonance. A good example of resonance is when singers have gifted sinus anatomy so that their singing voice resonates through the sinus cavities and forms very pleasant harmonics (discussed in chapter 15). These are individuals with deep rich singing voices, or much sought-after announcers. With unique details of our pharynx, sinus anatomy, and vocal cords, we have unique voices. God made some to be singers and/or speakers. Some voices have an encouraging sound to them. These are examples of spiritual gifts given to different members of the Body of Christ.

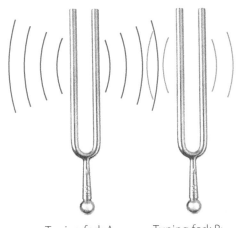

Tuning fork A: Vibration

Tuning fork B: Resonating with sympathetic vibrations

An interesting phenomenon occurs if you are near railroad tracks when a train is coming. The whistle from the approaching train appears to go up in pitch as the train approaches and down in pitch as the train moves away. **Pitch** is our perception of sound that is caused by a change in frequency. As the train approaches, the distance to the train gets shorter and shorter, so the sound waves from the train whistle have less and less distance to travel. This causes the areas of condensation to reach you sooner, which makes the frequency go up. As the train is moving away from you, it takes longer for the areas of condensation to reach you. With fewer areas of condensation reaching you in the same amount of time, the frequency goes down. The actual frequency of areas of condensation given off by the train whistle does not change. The velocity of the sound wave from the whistle also does not change. But you still perceive the change in the frequency of the whistle because of the motion of the train. If the source of the whistle was stationary and you moved toward and then away from it, the same thing would happen. This perception of the change in the frequency of sound is called the **Doppler Effect**.

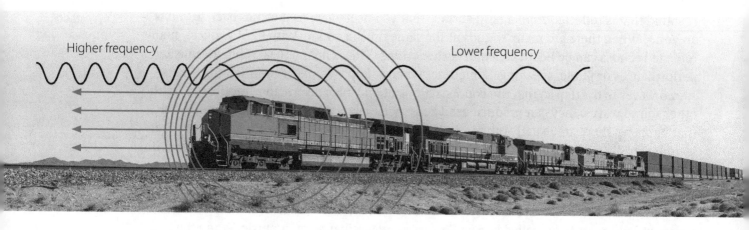

When motion and position cause changes in the perception and measurement of phenomena, the concept of a **frame of reference** is used. When you are in a jet going over 500 miles an hour and you get up to use the restroom, you feel as though you are standing still (except for sudden turbulence). Your frame of reference is the inside of the airplane, so, relative to the airplane, you are not going over 500 miles an hour. When you change your frame of reference to the ground, you are going over 500 miles an hour. If you moved away from the train at the same rate that the train moved toward you, there would be no Doppler Effect occurring even though both you and the train are moving.

If you are going down a highway alongside the railroad tracks, going faster than a train next to you, is the train going backward? What happens if you go faster than the train and catch up to it and pass it while the engineer is blowing the whistle? As you approach the train, the frequency of the whistle will increase and then it will decrease when you go past and beyond the train.

What is the loudness of sound? It depends upon whether you are measuring it with a meter that measures decibels or your hearing. The same sound can be annoyingly loud to a person with sensitive hearing and just right or even too low to someone with hearing difficulties. If you measure it with a meter, you get an objective measurement based upon the intensity of the sound. Meters do not need hearing aids. **Intensity** is the pressure (force per unit area — like pounds per square inch) of the condensations. Decibels, with the term "bel" named for Alexander Graham Bell, who invented the telephone, are one-tenth (deci-) of a bel. Hearing, however, depends upon the frequency of sound. You can usually tell if something is loud or not, but we distinguish between different sounds by their frequencies. How loud does a sound have to be for you to hear it? This varies with the frequency of sound. We hear some frequencies better than others. We have a threshold of hearing for different frequencies, which is the minimum loudness required to hear it. The word *loudness* is used here because that is what we perceive. Intensity is what a meter perceives. We also have a threshold of discomfort for different frequencies. This can vary between individuals, depending upon the sensitivity to each frequency. Music and sound are very subjective. Some music can be worshipful to some and near the threshold of discomfort for others. Each person has some frequencies that he/she is more sensitive to and some that are hard to hear. The pattern of frequency sensitivities can vary, causing sound to sound different to different people.

The standards for measurable decibels (db) are based upon hearing, where 0 decibels are the minimum intensity that can be heard. It is usually between 16 and 20 Hz frequency. Some average decibels are. . .

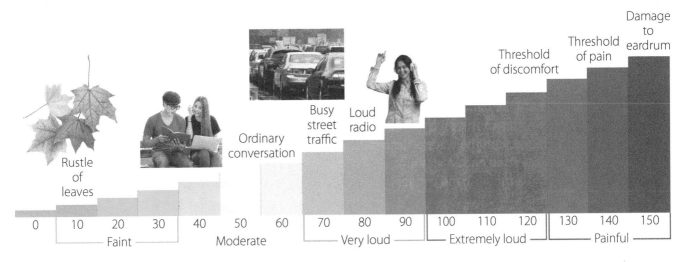

We must be careful as we go through life because we can gradually lose our hearing at important frequencies without realizing it. We do not have to reach 160 db to weaken the structures in our inner ear that become more susceptible to permanent damage.

It is a blessing that God gave us brains to interpret the sounds that we perceive. Again, this is a case where science is a blessing when kept in perspective. In science we continue to learn and adjust our understanding, while Scripture never changes, even though we still have much more to learn.

LABORATORY 14

Sound Waves

REQUIRED MATERIALS
- Rubber band
- Empty 2 liter soda bottle

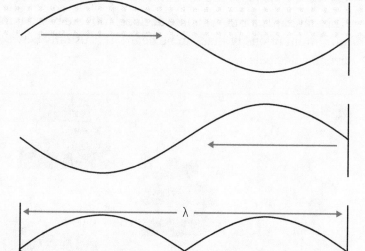

Introduction
A standing wave can be produced in the rubber band from the wave traveling down the rubber band combined with the wave reflected from the fixed end. A standing wave in a rubber band can look like this.

A standing wave in a 2 liter soda bottle is also where a wave and a reflected wave combine to produce a louder sound. As water is added to the bottle, the standing wave has less room and is shortened. This diagram shows standing waves of $\frac{1}{4}$ of a wavelength ($\frac{\lambda}{4}$), $\frac{3}{4}$ of a wavelength ($\frac{3}{4}L$) and $\frac{5}{4}$ of a wavelength ($\frac{5}{4}L$).

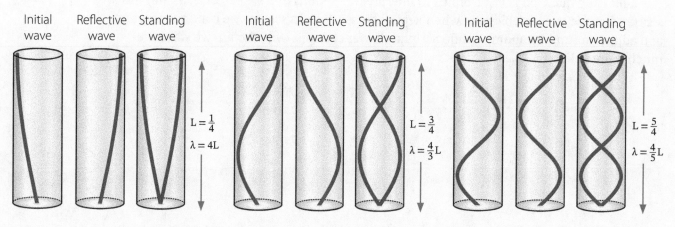

The sounds are loud when an area of condensation reaches your ear at the top of the bottle. This is where the lines are far apart (the antinodes).

Purpose

This exercise is to provide experience in producing standing waves in a stretched-out rubber band and in the air in a 2 liter soda bottle. This is the underlying principle in the design of musical instruments.

Procedure

 observe

1. Cut a rubber band so that you have a long rubber strip. Tie one end of the rubber band to a secure object on a wall. Pull the free end of the rubber band and give it an up and down pulse that travels down the rubber band so that it reflects at the fixed end and travels back toward you. Describe the wave as it travels down the rubber band and comes back toward you. Diagram the wave motion. (*Diagram L14.1*)

2. Redo step 1 but send a second wave that intersects with the first one after it reflects toward you.

 question

3. Describe what happens as the waves intersect with each other. This is like wave patterns of sound.

 research

4. Measure the length of the empty 2 liter soda bottle from the top to the bottom. Blow over the top of the soda bottle until you can hear a note (a standing sound wave). Notice how loud it is. Add an inch of water to the bottle and do it again. Make a chart showing how high the water is in the bottle and what it sounds like when you blow over its' top. Write how loud it sounds to you. Add another inch of water to the bottle and do it again. Keep doing this until you can no longer make a sound in the bottle.

 hypothesis

5. How full did it get at that point? At what depths in the bottle was the sound the loudest? These are the depths where you get resonance from the standing waves in the bottle.

 experiment

6. From this data, determine the wavelength of your sound.

 analyze

7. How many points of louder sound did you find in the soda bottle?

 conclusion

8. The less water you have in the bottle, the faster the bottle vibrates, creating a higher-pitched sound. The more water added in the bottle, the slower the bottle vibrates, which creates a lower pitch.

CHAPTER FIFTEEN
MUSICAL SOUND WAVES

OBJECTIVES

At the conclusion of this lesson the student should have an understanding of

- Music as condensations and rarefactions of air
- Subjective and objective analysis of music
- Pitch
- Musical scales
- The tuning of musical instruments
- Loudness and intensity
- Quality (timbre)

Music is a very broad subjective subject. Would you call the sound of someone lecturing music? Probably not. For our purposes, music is approached in this chapter as repeating patterns of condensations and rarefactions. There is much more repetition in music than in lecturing, even though both are supposed to be pleasant sounds in contrast to noise.

Noise or music is a matter of our brain's interpretation of the patterns of sounds that vibrate our eardrums. Our brains have a remarkable ability to distinguish patterns in sound waves. Sound waves that do not have patterns are noise. That is frustrating, as our brains search for patterns.

Neural stimulations from our inner ears are carried to several parts of our brains where multiple interpretations take place. Some go the areas that give us the sensation of sound and some go to the memory regions to pull up memories of the same sounds heard before. Perhaps a song was the same one that you heard when you first met a very special young man or young lady. Or maybe you heard that song in a very frightening situation that brought up those feelings of fear again. Music is very powerful with our emotions and we must be careful that they do not fool

Noise

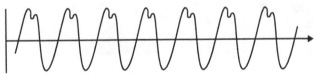
Music

us into making wrong decisions. We need to rest in the security of Christ rather than giving in again to fear. Perhaps the emotional appeal of music can make you feel like making a premature decision. Also, some people can express their feelings through music, which can be a blessing.

The study of the physics of music can seem to be very sterile and not do it justice. Music goes far beyond what we can do in a physics study. These principles are a foundation on which a rich history and resource is developed. Music is a gift from God that enables us to worship Him, celebrate important events, and to enjoy each other's company. Beyond the physics of sound waves is the capacity that God has given us, created in His image, to be able to "hear" music in our heads and create new pieces of music and arrange them for the diverse forms of instruments in an orchestra. Are we going to have music in heaven? Most certainly — in its purest richest form.

Analyzing music is usually a combination of using instruments that are more objective rather than perceptions that can be emotional. Music is usually analyzed for pitch, loudness, and quality (timbre).

Pitch is the frequency of the condensations. It refers to the number of crests in a wave that pass a certain point every second. In singing, pitch is varied by muscles stretching or relaxing the vocal cords. We distinguish different notes by their frequencies. In western cultures, the diatonic C major scale is a standard.

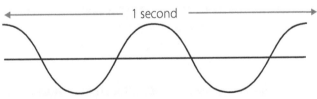
Low frequency — Low pitch — Low sound

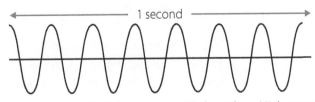
High frequency — High pitch — High sound

Note	Frequency
C	264 Hz
D	297 Hz
E	330 Hz
F	352 Hz
G	396 Hz
A	440 Hz
B	495 Hz
C	528 Hz

C at 528 Hz has twice the frequency as C at 264 Hz and is an octave higher. The word octave comes from Latin for 8 because it is 8 notes higher.

When string instruments are tuned, the tension on the strings can be varied to increase or decrease the frequency. Varied lengths and thicknesses of the strings can also vary the frequencies, giving different notes. Reed and brass instruments are tuned be varying the length of the vibrating air column in the instrument. Different notes can be played by letting air out at different places along the tubing of the instrument which varies the length of the vibrating air column.

Loudness is the brain's interpretation of the amplitude of sound. **Intensity** is an objective measurement. As part of a sound wave, intensity refers to the square of the amplitude of the wave. Loudness also depends upon an individual's sensitivity to sound. Someone who hears fainter sounds will consider music too loud, whereas someone else might consider it to be just right. Sensitivity to loudness varies as well with different frequencies because we hear some frequencies better than others.

The decibel scale is used to measure loudness. The faintest sound that can be perceived is 0 db. The db scale is a logarithmic scale where 20 db is 100 times (10^2) louder than 0 db. The exponent of 2 comes from the 2 in 20 db; 60 db (ordinary conversation) is 1 million (10^6) times louder than 0 db; 80 db

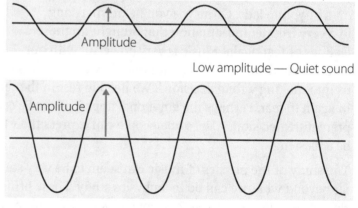

Low amplitude — Quiet sound

High amplitude — Loud sound

(loud radio) is 100 times (10^2) louder than 60 db (conversation). Our perception and actual loudness may be two different things. For example, a loud radio may not seem to come close to being 100 times louder than ordinary conversation. But the effect of the loudness on the structures of the inner ear do not depend upon our perception, but rather upon the power used to produce the sound. A person can gradually lose hearing ability without realizing it, just as a chicken can be boiled in a pot by gradually increasing the heat. Protect your ears because the only recourses are hearing aids and surgery. Damaged inner ear structures do not regenerate.

Another very important part of music is **quality** (or **timbre**). This is the pleasant quality that we associate with music. It involves **fundamental tones** and their **overtones** (also called **harmonics**). This can be easily seen in a guitar or violin. Consider the waves shown below to be the motion of the strings of a guitar or violin.

The number and loudness of the harmonics give the quality to music. Our brains have a remarkable ability to distinguish between the different harmonics in music, even though we may not even be aware of it. In music, these are called singing parts.

In string instruments, the strings vibrate and send sound waves into the instrument to the sound board which vibrates and produces resonating harmonics that give a richness to the sound. This requires careful selection of the wood from which the instrument is made. The sound board absorbs quite a bit of wave energy, so string instruments are limited in their volume. This is why there are usually many more string instruments in an orchestra than other instruments.

Some singers are blessed with good pharynx anatomies so that the air that passes from the lungs through the vocal cords resonates in the sinus cavities, producing rich harmonics.

In 530 B.C., Pythagoras noticed that stringed instruments were pleasing to the ear when the frequencies were ratios of whole numbers (overtones). He found that the strings of string instruments could be made to vibrate at different harmonics. He used the motion of the strings and the sounds they produced. This was long before any tuning instruments were available. Notes in a scale have frequencies that are whole number multiples of each other. The differences in frequencies between notes in a scale are called **intervals**.

Intervals are called **melodic** if one comes right after the other in a melody, or they are vertical or **harmonic** if they are at the same time such as in a **chord**.

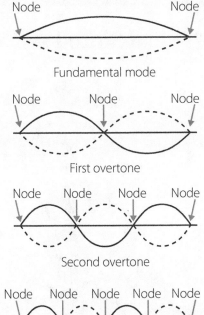

Fundamental and overtones (harmonics)

The differences in octaves of the same notes are multiples of 2. In the C scale given above, the higher C (528 Hz) is twice the frequency of C (264 Hz) an octave lower.

In 1822, the French mathematician Joseph Fourier devised a mathematical method to divide wave form that is a combination of wave patters into its fundamental and harmonic overtones. Your brain does this automatically when you hear music. When you sing into a microphone that is connected to an oscilloscope (converts condensations in air into electric signals), you can see that composite waveform of your voice.

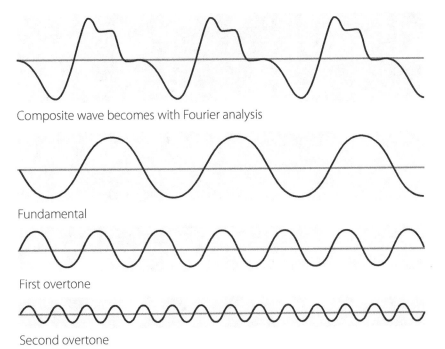

Sound is the movement of pulses of compressed air molecules that we perceive as hearing when they cause vibrations of our eardrums. Music is composed of pulses of compressed air in specific patterns. The vibrations of our eardrums cause ear bones in our middle ear to further cause vibrations in the fluids of the inner ear. Pulses of waves in this fluid stimulate nerves that carry impulses to our brains that are interpreted as sound.

LABORATORY 15

Musical Sound Waves

REQUIRED MATERIALS

- A musical instrument and someone to explain how it is tuned. This can be your own, borrowed, or seen at church or a music store.

Introduction

In an orchestra, it is necessary to tune the instruments so that the same notes have the same frequencies to avoid beats of loud sounds between softer sounds. This is accomplished by adjusting the tension in string instruments, the length of the vibrating air columns in woodwind and brass instruments. Percussion instruments are usually tuned by adjusting the tensions on the vibrating membranes.

Purpose

The purpose of this exercise is to gain experience in tuning an instrument to produce the same frequency as another instrument and how it produces overtones (harmonics).

Procedure

This exercise is very different from the others. You are to use a musical instrument to learn how to tune it and how it produces overtones. You should have an experienced person explain and demonstrate to you how to tune their instrument and then demonstrate what it sounds like along with the instrument to which it was tuned. Also, discuss how resonance produces overtones in this instrument. Be sure to explain what overtones (harmonics) are. You are to write a one and a half page double spaced, 12 font, paper describing how it was done and what it means to tune an instrument. Any pictures or diagrams are not part of the one and a half pages. This may be your own instrument and from your own background.

CHAPTER SIXTEEN

ELECTROMAGNETISM

OBJECTIVES

At the conclusion of this lesson the student should have an understanding of

- Electric charges
- Electric force fields
- Electric potential
- Electromagnetism
- Electric monopoles and magnetic dipoles
- Magnetic domains
- Earth's magnetic field
- Electromagnetic motors
- Electromagnetic induction of electric current

Chapter 16 introduces electric and magnetic force fields which lead up to the understanding of light as introduced in chapter 17.

By the 1700s, it was apparent that objects could be "charged," and that there were basically two different charges. This is what happens when you drag your feet on the carpet on a warm, dry summer day and get zapped when you touch a light switch (or sleeping cat). Or you brush your hair and it sticks up in all directions. An example is when you run a nylon comb through your hair and bring it near a very narrow stream of water from a faucet. The stream of water moves away from the comb indicating that the comb and the water have like charges. It was noticed that a rubber rod rubbed on cat fur attracted a glass rod rubbed on silk, and also that rubber rods rubbed on cat fur repelled each other and two glass rods repelled each other.

Benjamin Franklin (1706–1790) proposed in 1750 that when an object had an excess of "charging fluid" it was positively charged, and when it lacked this fluid, it was negative. He was not clear as to which objects were positive and which were negative because he could not see this fluid. Later, the English physicist J.J. Thomson published his discovery of the electron in 1897, where he identified it as negatively charged after the identification of charges by Benjamin Franklin.

In the 1780s, Charles Coulomb conducted experiments to show that the force between two charges could be expressed by the equation . . .

$$F_e = \frac{kq_1q_2}{r^2}$$

where F_e is the force between the charges, k is a constant (called Coulomb's constant), q_1 and q_2 are the charges, and r is the distance between the charges. Do you see the similarity between this equation and that of the Law of Gravity expressed by Newton as described in chapter 5?

$$F = \frac{(Gm_1m_2)}{d^2}$$

This is a pattern that is often seen in the created universe. They both obey the Inverse Square Law where the force spreads out over distance. If you double the distance between the objects, the force between them is reduced by 4 times (2^2). A major difference between them is that gravity is a monopole (it is only expressed in one direction). There is no negative gravity. To have such, there would have to be negative mass which does not exist. Electric charges are dipoles, both positive and negative, so they can be positive or negative.

The concept of an **electric force field** was published in 1865 by James Clerk Maxwell. He said that there was no invisible fluid, as expressed by Benjamin Franklin, but that the objects affected the space around them. He drew a diagram for electric force fields like the vector diagram in chapter 5 for gravity force fields (*Diagram 16.1*).

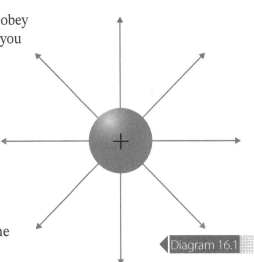

Diagram 16.1

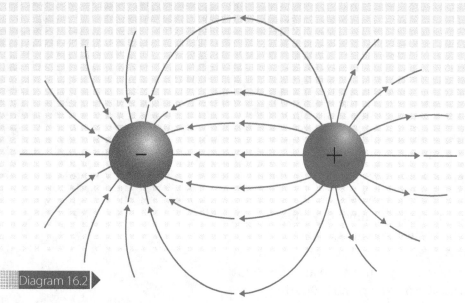

Diagram 16.2

This idea had been informally introduced earlier by Michael Faraday (1791–1867) as field lines to describe electric and magnetic effects.

The field lines in the diagram provide a way to visualize the direction and strength of the field. It is especially important for electric field lines because they can be either positively or negatively charged. Maxwell had the field lines coming from positive charges and ending on negative charges.

Recall from the study of gravitational potential energy that the potential energy of an object in a gravitational force field depended upon the height of the object.

$$PE = mgh$$

An object 5 meters above the ground would have more potential energy than an object 1 meter above the ground. When the object fell from 5 meters, it would have a greater kinetic energy right before it hit the ground than the object falling from 1 meter.

If you move a positive charged object close to a positive charged sphere, there will be a repulsion between the objects. The closer you get with the positively charge object, the greater the repulsion and greater the potential energy. It will fly away with more energy when released from a closer position. The gravitational field around earth depends upon the mass of earth, and the charged sphere has a force field that depends upon the amount of charge in the sphere. The gravitational potential energy depends upon the gravitational field of earth, and the potential energy of an object depends upon its mass. In an electric force field, the force field depends upon the charge of the source (sphere in this example) and the potential energy of an object in the force field. This potential energy is called the **electric potential**. If an object in a gravitational force field has more mass, its potential energy is greater. If a charged object in an electric field has more charge, it will have more electric potential energy. For this reason, it is traditional to speak of the electric potential as the amount of energy (potential energy) as energy per charge or . . .

$$\text{Electric Potential} = \frac{\text{energy}}{\text{charge}} = \frac{\text{joules}}{\text{Coulomb}} = \text{volts}$$

The electric potential is measured in units of voltage or **volts**. The energy is in units of joules and the charge is in units of Coulombs of charge, which is the charge of 6.25×10^{18} electrons (negative) or protons (positive). This is 6,250,000,000,000,000,000 electrons. This amount of charge is called a **Coulomb** and is the value used for electric potential.

Raw magnetite (lodestone) crystals

The ancient Greeks discovered stones (called lodestones) that had unusual properties in that they attracted pieces of iron. They found these stones on the island of Magnesia, so they called them magnets. In the 1500s, William Gilbert, Queen Elizabeth's physician, found that he could make iron magnetic by rubbing it against lodestone. In 1820, a Danish high school physics teacher, Hans Christian Oersted, found during a classroom demonstration that electric currents affect magnets. When Maxwell described the electric force fields as affecting the space around electric charges, it was realized that magnets also had force fields around them. It was found that electric forces and magnetic forces were related when they observed moving electric charges producing **magnetic force fields**. These were two different aspects of the same thing, called **electromagnetism**.

Electric charges are called **monopoles** in that an object is either positive or negative, not both. Magnets are **dipoles** in that they always have two poles called north and south. Wires with electric currents are surrounded by magnetic fields. What about solid magnets that have no electric current? Why are they magnetic? It was later found that negative-charged electrons within atoms move about the nucleus and spin. This motion produces magnetic fields. Why is not everything magnetic, because all atoms have moving and spinning electrons? In most atoms, electrons are paired and spinning in opposite directions. This way their magnetic fields are opposite to each other and cancel each other out. However, the atoms of iron, nickel, and cobalt have some unpaired electrons that do not cancel out each other.

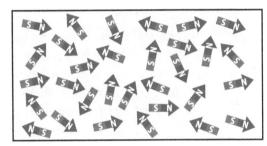

Domains before magnetization — random direction

It was found that just as opposite electric charges attract each other, and like electric charges repel each other, opposite magnetic poles attract each other and like magnetic poles repel each other.

The magnetic fields of iron atoms are very strong so that they line up in groups called **domains**. When the domains line up with each other so that their north and south poles are facing the same direction, the larger piece of iron is magnetic. As you know, most pieces of iron are not magnetic. This is because their domains are not lined up with each other. This is like a bunch of little magnets facing different directions. But when William Gilbert rubbed a piece of iron with a magnet, the magnet caused the domains to align with each other, producing a larger magnet. Several years ago, when I was working on an electric lawn mower, I used

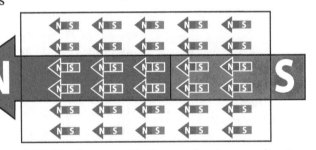

Domains after magnetization — align in same direction

a screwdriver with a long iron blade to remove some screws. As I pushed the blade of the screwdriver down near a very strong magnet in the lawn mower's motor, the blade of the screwdriver was grabbed by the magnet and was removed with great difficulty. From that time on, the blade of the screwdriver was a very strong magnet. It is interesting that when a permanent iron magnet is dropped or heated, its magnetic field is weaker. The domains of iron atoms are disrupted and not aligned together as well as before.

Earth is a large magnet with a north and south pole. Magnets in compasses line up with the magnetic field of the earth. The pole pointing toward earth's north is called the north seeking pole and the pole facing earth's south is called the south seeking pole. For simplicity's sake, the north seeking pole is called the north pole and the south seeking pole is called the south pole.

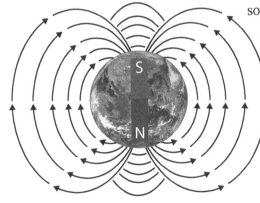

Earth's magnetic poles are not the same as earth's geographic poles. The north pole of earth's magnet is near earth's southern geographic pole, and earth's magnetic south pole is near earth's geographic north pole. The north pole of a magnet in a compass points toward the magnetic south pole of earth in the general direction of the geographical north pole.

The magnetic field of earth is thought to be formed by moving ions (charged atoms and molecules) in convection currents in the molten outer core below earth's mantle. As molten iron deposits in volcanic lava cool, they form solid rock. The magnetic domains in the rock align with the magnetic field of the earth before they solidify. These indicate the direction of earth's magnetic field when they solidified. Along mid-Atlantic Ocean ridges, molten volcanic lava comes up to the ridge and solidifies, spreading the edges of the ridge farther apart. The rock on both sides of the ridge have sections of rock whose magnetic domains point in opposite directions, indicating that perhaps the magnetic field of the earth reversed itself at different times. It has been assumed that this would take a long time. This has been cited as evidence for a very old earth. It was suggested recently that if the reversal were rapid, there would be a very thin section of reversed domains in solidified rock. This was found in 1989 in solidified lava flow in Steens Mountain, Oregon. This would be consistent with a young earth model.

The magnetic field around earth is like that of a huge bar magnet. This magnetic force field traps most of the deadly cosmic rays coming from the sun and other stars. It is named after its discoverer — the Van Allen radiation belts. If the magnetic field of the earth had reversed over long periods of time, there would have been a time period with no magnetic field around earth to capture cosmic rays. This would have been devastating to life forms on earth at that time. This also favors a young earth model.

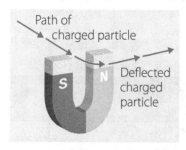

It was found that a charged particle at rest does not interact with a non-moving magnetic field. But if a charged particle is moving perpendicular to a magnetic force field, it will experience a deflecting force. This was the demonstration conducted by Hans Christian Oersted in 1820 in his high school class.

If the wire is a loop and the electric current through the wire is reversed back and forth, the wire can be caused to rotate within a permanent magnet. This is an **electromagnetic motor**. An unexpected observation in Oersted's classroom led the way to the development of an electric motor (*Diagram 16.3*).

When an electric current flows through a wire, magnetic field lines exist as circles around the wire. The direction of the induced magnetic field can be shown by grasping the wire with your right hand with your thumb pointing in the direction of the current flow and wrapping the fingers of your right hand around the wire indicating the direction of the magnetic field lines. This is called the right-hand rule.

◁ Diagram 16.3

When a wire is wrapped around a piece of iron (like an iron nail) with an electric current flowing through the wire, the magnetic domains in the iron line up, producing a strong magnet. This is an **electromagnet**.

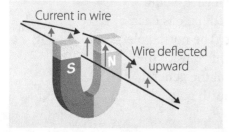

It was also found that when a magnet was passed through a wire coil, an electric current was induced in the wire. This led to the development of electric generators and alternators. This was discovered independently by Joseph Henry in America and Michael Faraday in Scotland in 1831. This was **electromagnetic induction** that was a breakthrough in the generation of electricity.

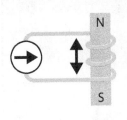

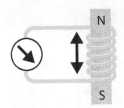

Twice the voltage induced in the wire

The electric generator and motor greatly changed the way of life.

Another application of electromagnetic induction that has greatly affected our way of life is a **transformer**. In this case, an electric current in one wire produces a magnetic field that induces an electric current in another wire. This is used to increase or decrease the voltage of an electric current. Electric current is carried at much higher voltage in the main power lines that lead to your home because it is more efficient for traveling over great distances. Before the current can go to your home, it must be reduced to safer voltages. A transformer has a primary coil which is wire coiled around an iron core. It produces a magnetic field that induces an electric current in a secondary coil of wire also wrapped around the iron core. In the following diagram, the primary coil has 12 loops around the iron core and the secondary coil has 4 loops around the iron core. The wire of the primary coil does not touch the wire of the secondary coil (*Diagram 16.4*).

If 300 volts is carried in the primary coil, there will be 100 volts in the secondary coil. This is because the number of loops is 3 to 1 between the primary coil and the secondary coil. There is less wire exposed to the magnetic field produced by the primary coil in the secondary coil. Did you lose energy in going from 300 V to 100 V? Not really. Remember back in chapter 7 how the work done with a machine required less force but made up for it by having to exert the force over a greater distance? In this case, the voltage is reduced by $\frac{1}{3}$ in the secondary coil, but the amount of current (amount of electron flow) increased by 3 times as much. This is called a step-down transformer because the voltage "stepped down." Consider this transformer which is a step-up transformer (*Diagram 16.5*).

In this case, the loops in the secondary coil are 3 times as many as in the primary coil. There are 100 V in the primary coil and 300 V in the secondary coil. Again, there is $\frac{1}{3}$ as much current in the secondary coil. An important example of a step-up transformer is the coil in the starter of an automobile. It takes a lot more voltage than the 12 volts from a car battery to start a car.

Faraday stated (Faraday's Law) that an electric field is induced whenever a magnetic field changes with time. Maxwell confirmed the inverse of Faraday's Law by stating that a magnetic field is induced whenever an electric field changes with time. As you will see in the next chapter, these two statements form the basis for understanding light — electromagnetic radiation.

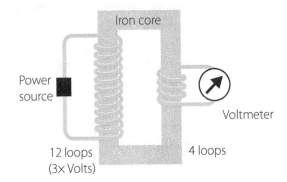

Step-down transformer

Diagram 16.4

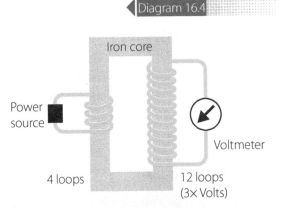

Step-up transformer

Diagram 16.5

LABORATORY 16

Electromagnetism

REQUIRED MATERIALS

- Nylon comb
- Bar magnet
- Iron filings
- Iron nail
- Insulated wire
- 2 × AA batteries
- AA battery holder

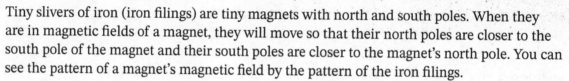

Introduction

When atoms or molecules have more negative electrons than positive protons, they have an overall negative charge. When atoms or molecules have fewer negative electrons than positive protons, they have an overall positive charge. Objects with opposite charges attract each other and objects with like charges repel each other.

Tiny slivers of iron (iron filings) are tiny magnets with north and south poles. When they are in magnetic fields of a magnet, they will move so that their north poles are closer to the south pole of the magnet and their south poles are closer to the magnet's north pole. You can see the pattern of a magnet's magnetic field by the pattern of the iron filings.

When charges move through a conductor (wire), a magnetic field is induced around the wire. This is the inverse of Faraday's Law as Maxwell proposed.

Purpose

This exercise demonstrates the attraction of opposite charges and the repulsion of like charges; the pattern of magnetic field lines of a bar magnet and the effects of a magnetic field induced by moving charges.

Procedure

 observe

1. Take a nylon comb and run it through your hair then hold it over some tiny pieces of torn paper.

 question

2. How would you describe what happens?

research

3. Run the comb through your hair again and hold it to the side of a narrow flow of water from a faucet. Describe what happens to the stream of water. In both cases, you are looking at the attraction of opposite charges and the repulsion of like charges. These charges are called static electricity because the electrons are not flowing through a conductor.

4. Lay a sheet of paper over a bar magnet. Sketch the shape of the bar magnet on the top of the paper. Sprinkle iron filings onto the paper and notice how they line up with the magnetic field lines of the bar magnet. Make a drawing on another sheet of paper of the bar magnet and the pattern of the iron filings. Indicate which is the north pole and the south pole of the magnet.

5. Wrap wire around an iron nail leaving about $\frac{1}{2}$ inch free at the tip of the nail. Connect both ends of the wire to a battery holder and place 2 AA batteries in the battery holder. Keep the battery connected only as long as you need so that you do not drain the battery and it does not get too warm. Bring the tip of the nail to a pile of iron filings. Describe what happens. Clean the iron filings off the nail. Place the nail on a tabletop and cover it with a sheet of paper. Sprinkle iron filings on the paper over the nail. Describe the pattern of the iron filings.

hypothesis

6. What can you say about the pattern of the magnetic field lines of the magnetic nail?

experiment

7. Prepare a second nail with a wire wrapper around it like the first one. Connect the ends of the wire to a second battery holder with 2 AA batteries. Bring the ends of the two nails near each other.

analyze

8. How do they react to each other? Place both nails under a sheet of paper and sprinkle iron filings onto the paper? Draw the two nails underneath the paper and the pattern of iron filings on top of the paper.

conclusion

9. What can you say about the magnetic fields produced by your electromagnets?

CHAPTER SEVENTEEN

LIGHT WAVES – ELECTROMAGNETIC RADIATION

OBJECTIVES

At the conclusion of this lesson the student should have an understanding of

- Light created by God
- Light as electromagnetic radiation
- Only one velocity for light
- Frequencies and wavelengths of light
- Light propagation
- Electromagnetic spectrum
- Emission spectra
- Wave and particle models of light

The spectacular aurora borealis, or the "northern lights," over Canada is sighted from the space station near the highest point of its orbital path.

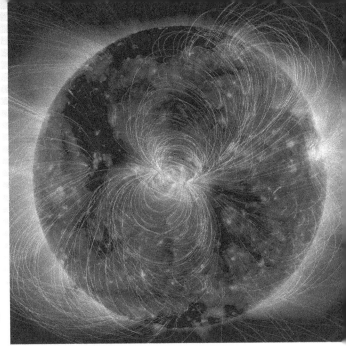

In the beginning God created the heavens and the earth. The earth was without form, and void; and darkness was on the face of the deep. And the Spirit of God was hovering over the face of the waters. Then God said, "let there be light"; and there was light. And God saw the light, that it was good; and God divided the light from the darkness. God called the light Day, and the darkness He called Night. So the evening and the morning were the first day (Genesis 1:1–5).

The sun, moon, and stars were not created until the fourth day. How then could there be light on the first day? Some have suggested that the days were not meant to be separate 24-hour days. But the text states that evening and morning separated them. Some have suggested that this text is not literal but an allegory. However, the grammatical form in the Hebrew is that which is used for literal history. Even a Hebrew youth would recognize that. This idea of light from God without the sun comes up elsewhere in Scripture. In Revelation, John writes that Jesus gave him a vision of the new Jerusalem where "The city had no need of the sun or of the moon to shine in it, for the glory of God illuminated it. The Lamb is its light. And the nations of those who are saved shall walk in its light, and the kings of the earth bring their glory and honor into it" (Revelation 21: 23–24).

In Genesis and in Revelation the references are to literal light that you can see by. Can you have light without a source — something to emit the light? If God creates it — yes. Would this light be the same as we see by today? Genesis 1 distinguishes day from night and in Revelation it is the light that people will see by. Since the days of Adam and Eve, light has guided the affairs of living creatures. Instead of trying to understand Genesis with minimum impact from God, it was written to express the maximum impact of God. God created the sun, moon, and stars out of love and grace because He knew that very shortly, those created in His image that He dearly loved would rebel against the glory of God and need to be restored. Light is such a gift, despite all the rebellion God has been subjected to. We were given eyes to see our crucified and risen Savior. God enabled one of His own, James Clerk Maxwell (1831–1879) in 1865 to describe light as a combination of magnetic and electric force fields.

Faraday stated (Faraday's Law) that an electric field is induced whenever a magnetic field changes with time. Maxwell confirmed the inverse of Faraday's Law by stating that a magnetic field is induced whenever an electric field changes with time. These concepts were developed in chapter 16 in preparation for developing an understanding of the nature of electromagnetic radiation.

A changing magnetic field induces a changing electric field, which induces another changing magnetic field, which induces another changing electric field, and on it goes. These new changing electric fields and changing magnetic fields keep moving out from the source. This is **electromagnetic radiation,** or as commonly called, light. The changing magnetic fields are perpendicular to the changing electric fields. This was a major clue that they were different force fields (*Diagram 17.1 on the following page*).

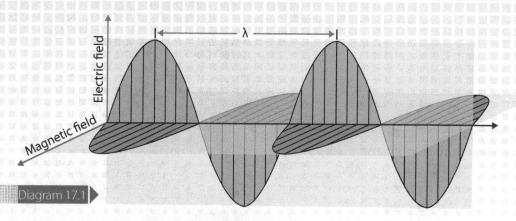

Diagram 17.1

It was found that the frequencies of these force fields differed depending upon their source, but their velocities were always the same. James Clerk Maxwell realized that electromagnetic radiation was light when he did some fantastic mathematical gymnastics to calculate the velocity of the changing magnetic and electric fields and found that their velocity was equal to the velocity of light. That was the connection between electromagnetic radiation and light. This velocity was 186,411 miles per second or 300,000 kilometers per second. This supported the answer to one of the greatest questions in physics — the understanding of the nature of light.

James Clerk Maxwell and Katherine Maxwell, 1869

There is a story that the evening after he came to this realization, James had a date with his bride-to-be. She told him about the beauty and wonder of the stars and he told her that he was the only man on earth who understood the nature of the light from the stars and he was probably right. It would be interesting to know if she was as thrilled with that comment as he was. He realized that light was composed of electric and magnetic fields that kept inducing each other, spreading out without losing energy. Fortunately, he was a Christian man who gave glory to God for all that He created and revealed to us. Now you can see why so much time was spent on magnetic fields and electric fields inducing each other in chapter 16.

Maxwell calculated that if the velocities of the induced fields were below a certain value, they would gradually die off and energy would be lost. This is against the Law of Conservation of Energy so he knew that that could not be correct. If the velocities of the force fields were above a certain value, they would keep increasing in energy as they spread out. That could not be the case because the Law of Conservation of Energy also states that energy cannot come from nowhere. So, the Law of Conservation of Energy (energy cannot be created or destroyed) was the key to finding the velocity of electromagnetic radiation.

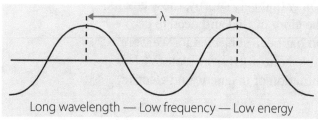

Long wavelength — Low frequency — Low energy

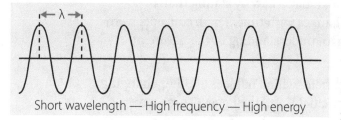

Short wavelength — High frequency — High energy

The electromagnetic force fields were found to vary in frequency and wavelength, but not in velocity. Each combination of induced magnetic and electric force fields that travel out from the source is treated as the crest of a wave and the spaces in between are the troughs of the waves. The distance between each group of force fields is the wavelength. The number of groups of force fields to pass a given point each second is the frequency of the wave. If the source of the force fields has more energy, more groups of force fields will pass in one second, meaning that it has more energy. Higher frequency and lower wavelengths mean more energy, and lower frequencies and longer wavelengths mean less energy.

In sound waves, the areas of condensation do not go all the way to their destination, but the magnetic and electric force fields do go all the way to their destination. Sound travels through a medium because something must be compressed to make an area of condensation, but light can go through a vacuum because the electric and magnetic force fields are not part of the medium through which they are traveling.

Radio waves have the lowest energy and gamma rays have the greatest energy. A vibrating electric charge induces the magnetic field that induces another electric field. The vibration rate of the electric charge determines the frequency of the magnetic and electric force fields produced. The source of electromagnetic radiation is called the source antenna and the destination is the receiving antenna. The electrons in the receiving antenna are caused to vibrate by the incoming magnetic fields at the rate of the electrons in the source antenna. The sensory neurons (rods and cones) in the retinas of your eyes act as receiving antennae.

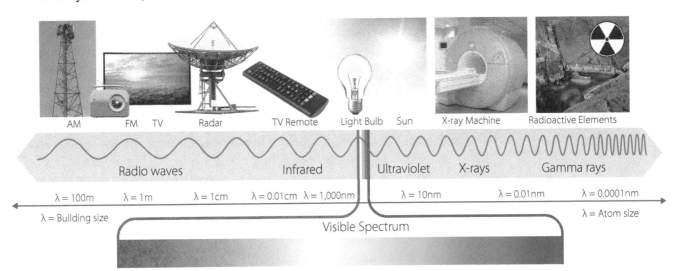

The lower energy electromagnetic radiation are radio waves, followed by microwaves and then infrared (lower than red). We cannot see any of these. Infrared radiations are called "heat waves" because we can feel their energy as heat. Infrared cameras can be used to measure the temperature of objects. Objects with higher temperatures emit more infrared radiation. These are followed by visible light. From lowest energy of visible light to higher energies of visible light are red, orange, yellow, green, blue, and violet. Electromagnetic radiation with energy greater than violet is ultraviolet followed by x-rays and gamma rays.

So far, the discussion has been about light traveling through a vacuum. What happens when light travels through a transparent medium like glass? Light takes longer to travel through glass than through a vacuum. Light only travels about 67% as fast through glass. Light energy is absorbed by the electrons in the silicon atoms in glass and re-emitted between the atoms and absorbed by the electrons of the next atom and so forth. Light travels at its normal velocity between silicon atoms and is eventually released at its original velocity after leaving. Therefore, light slows down in glass and is released at its normal velocity on the other side. The absorption and release of light by the electrons takes time, which increases the overall time that it takes for light to go from one side of glass to the other. Also, the intensity of light is diminished because some energy is released as heat in the glass. Light travels about 7% of its velocity through water. It is slower in diamonds — about 41% of its normal velocity.

How do microwaves work? In the electromagnetic radiation spectrum, the lower energy emission are the radio waves. The higher energy radio waves are microwaves. Visible light cannot pass through the walls of your house, but radio waves can. Therefore, you can listen to the radio inside your house. Microwaves can pass about a centimeter into food (your flesh too), which is why the door of your microwave oven must be secured for it to operate. The inside walls of the microwave oven bounce the microwaves around so that they reach your food. If the food is thick, the outer centimeter heats a lot more. The microwaves are absorbed by the electrons in the food, especially water, which increases their vibration rates, which raises their temperature. Sometimes the outer portion of the food, especially meat and other thick foods, can be much hotter than the interior. Therefore, the instructions state to let the food stand outside the oven for a period. This allows the hotter regions to heat the cooler parts by conduction. There is a lot of steam produced in microwaved food.

Microwaves are also used to transmit information over large distances, such as from satellites. The frequencies used are different from those used for cooking, which is why they do not cook you.

Electromagnetic radiation is all around us — especially radio waves and infrared. I heard a person say one time that the fillings in his teeth acted like radio wave antennae which set up vibrations in his skull so that he constantly heard a certain radio station. The natural vibration rates of the electrons in his fillings would have to match the frequencies of the radio waves. I did not take him seriously. This has been especially helpful with infrared radiation. Objects with higher temperatures give off more infrared radiation. Satellites can measure the temperatures of the sea surface and cloud tops for weather predicting using infrared cameras. Physicians can identify inflamed tissues this way. Very small leaks in air ducts in a home can also be detected this way.

Atoms have what are called **emission spectra**. Electrons in an atom absorb electromagnetic radiation as energy. Electrons have only certain levels of energy. That is like having a car that only goes 10, 20, 30, or 40 miles an hour with nothing in between. Of course, that would be silly for cars — but not for electrons. A hydrogen atom has 1 electron with set possible energy levels. Depending on how much energy the electron will receive, it will acquire energy bringing it to a higher level. Then the electron loses the gained energy in the form of light. The frequency of the released light indicates the amount of energy released. The electrons from many atoms release energy from the several allowable energy levels of the electrons. The light emitted has specific frequencies with nothing in between those values. This gives a spectrum that looks like the following.

Hydrogen emission spectrum

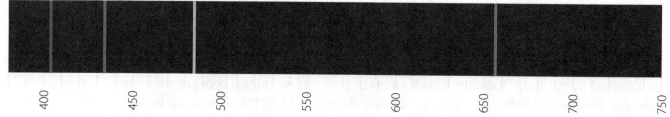

Light emitted from hydrogen will always give this pattern. The intensity of the light emitted indicates the relative amount of hydrogen present. By analyzing the light from distant stars, it was realized that stars consist of 90% hydrogen. The two electrons of a helium atom also give off a distinct emission spectrum. From that it was determined that stars are about 10% helium. They have other elements as well, but their emission spectra are very faint. How can you have more than 100%? The light from other elements fit into a margin allowed for experimental error.

Helium emission spectrum

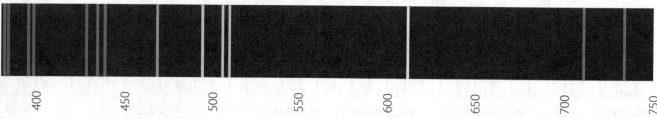

Helium got its name from the Greek word for the sun (*Helios*). When analyzing the emission spectrum of the sun, there were some emission lines that did not match those of any of the known elements. The strange pattern was assigned to an element in the sun that had been unknown on earth up to that time.

These emission patterns appear to be uniform for the entire universe. Such uniformity is certainly useful for us, but more so it points to a very intelligent plan in the mind of God who planned for us to have such a powerful tool. That is grace.

The chlorophylls and xanthophylls in plant cells absorb frequencies of light that eject electrons of atoms of these molecules that begins a chain reaction that results in the plant taking in CO_2 and H_2O to produce sugar ($C_6H_{12}O_6$) and O_2, which we so desperately need. That also is grace.

Light can be diffracted just like other waves. When light is diffracted through a barrier with two holes, interference patterns can be seen, as in the cases of sound and water waves. Therefore, light is often discussed as waves. This is called the wave model of light. A model is based upon the behavior of something rather than an actual physical appearance. An example is the atomic model. We know what an atom does, but we do not know what an atom looks like. This is especially true for electrons and protons in atoms.

Light also behaves as a beam of particles, instead of a wave. This is based upon the **photoelectric effect** of light. When light below a certain frequency shines on a solar panel, the panel does not gradually absorb energy from a wave constantly flowing onto it. Rather, electrons are only able to generate an electric current when the light is above a critical frequency. If energy in light were in a particle, the particle would have to have at least the minimum energy to eject an electron from an atom in the solar panel. Here, a model that describes light as consisting of particles (**photons**) of energy is used. Do these two models contradict each other? Yes. Are they both used? Yes. One model describes some behaviors of light and the other model describes other behaviors of light. The models do not say that light is a wave and a particle. They say that sometimes light behaves like a wave and sometimes like particles. God does not do anything in a simple way, does He? His creation is marvelously complex. But look at what a blessing the different frequencies of light are!

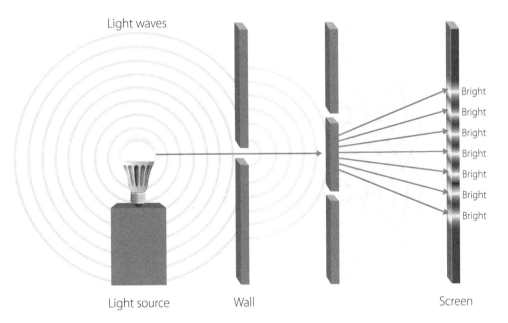

Light is electromagnetic radiation which is composed of electric and magnetic energy fields. Some of the properties of light are best explained by thinking of light as waves of energy and other properties are better explained by thinking of light as pulses of energy. Light and sight are a great blessing. It will be marvelous to see what it is like when Christ is the light that we live by in the new heavens and new earth.

LABORATORY 17

Light Waves – Electromagnetic Radiation

REQUIRED MATERIALS

- Insulated wire
- Bar magnet
- Galvanometer
- Glass of water
- Pencil
- Diffraction grating
- Single beam light source, pen light
- Glass of very diluted milk

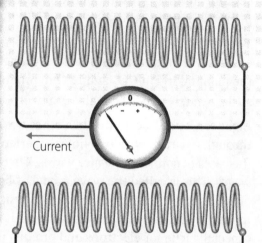

Introduction

A moving magnetic field induces a moving electric field and a moving electric field induces a moving magnetic field. Alternating magnetic and electric fields moving through space are electromagnetic radiation, which is light. It can vary in frequency and wavelength, but is always constant at the same velocity (indicated by the letter "c").

An electric current, which is a moving electric field, can be detected with a galvanometer. If the needle on the dial goes back and forth, the current is alternating in direction.

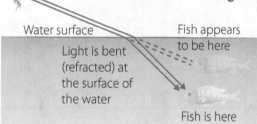

When light enters another medium, it appears to slow down and then go back to its original velocity when it passes out on the other side. This causes light to be refracted (bent) as it is going through water. When some of the light waves are going more slowly, they get behind the rest of the light waves. This causes the light waves to appear to be bent.

Light is diffracted (like you saw earlier with water waves) as it goes through a very narrow slit. When it spreads out, the colors are separated from each other. Violet and blue light have the greatest energy in the visible range and red has the lowest energy. The light waves with more energy are diffracted the greatest and those with less energy are diffracted less. The diffraction grating has many narrow slits producing a pattern of many diffracted light waves.

Purpose

This exercise demonstrates a moving magnetic field inducing a moving electric field, which is the essence of electromagnetic radiation, which is light. The refraction of light as some of the light waves go more slowly than the others is seen in a simple demonstration. Light waves are diffracted when they pass through a narrow slit, just like water waves. The diffraction pattern is shown that separates the different colors of white light.

Procedure

 observe

1. With a screwdriver, adjust the knob on the front so that the dial on the galvanometer is in the middle at the zero reading before you begin. Take about a couple of feet of insulated wire and form 20 loops by wrapping it around a round cylindrical object with about 2 inches of the wire exposed on both ends of the loops. Remove the wire from the object you wound it around. Connect the free ends of the wire to the posts on the galvanometer. Hold one of the bar magnets in the space between the loops.

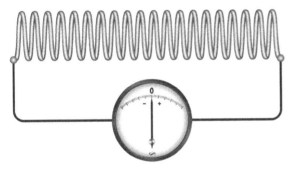

 question

2. Does anything happen on the dial of the galvanometer? This time pass the magnet back and forth through the space between the loops. Describe the response on the galvanometer. Is a moving magnetic field between the loops of wire inducing moving electric fields in the wire? How can you tell?

research

3. Many optical illusions are passed off as magic. In this procedure you are going to bend a pencil and then straighten it back out as good as new. Take a glass of

water and place a pencil about halfway into the water. Look at it from the side. Describe the appearance of the pencil with part of it in water and part of it out of water. When light slows down, it makes objects appear to be bent.

 hypothesis

4. What is happening to the light waves as they pass into the water and come back out on the other side?

 experiment

5. Place the diffraction grating upright on a tabletop with a piece of white paper next to it. Pass a beam of white light through a diffraction grating in a darkened room. The narrow beam of a pen light will work well for this. Position the white paper so that the colors are easily seen on the white paper.

 analyze

6. Describe and make a sketch of your observations. The diffraction of light supports the wave model of light.

7. Take a glass of water and add just enough milk to make it a bit murky. In a darkened room, with a penlight pass a light beam through the water. Describe what you observe in the glass. Place a sheet of paper on the other side of the glass and describe the light beam as it exits the water.

 conclusion

8. What different phenomena can you identify (refraction, diffraction, etc.)?

CHAPTER EIGHTEEN

LENSES – REFRACTION

OBJECTIVES

At the conclusion of this lesson the student should have an understanding of

- Wave fronts and light rays
- Refraction through flat glass
- Index of refraction
- Simple convex lens and focal point
- Diverging lenses
- Microscopes and telescopes
- A mirage

We are very fortunate to live in a time when there are multiple ways that poor eyesight can be corrected. There are so many aids for vision and means to measure our visual capabilities and lenses designed to make necessary corrections. God gave us our first lenses, our corneas, in the front of our eyes and the lens between the outer and inner eye chambers. The shape of our eyeballs and the viscosity (thickness) of the fluids in our eyeballs are also involved in focusing images upon our retina.

The key to understanding the functioning of a lens is **refraction**. That is the bending of light rays as they pass through media at different velocities. The crests of light waves (the areas of greatest concentration of electric and magnetic fields) can be drawn as **wave fronts**. Light rays are drawn perpendicular to wave fronts.

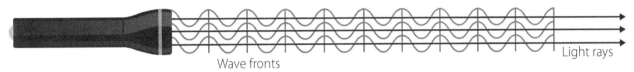
Wave fronts — Light rays

When light rays are perpendicular to the surface of a flat piece of glass, they appear closer together in the glass because they are going more slowly and farther apart after they leave the opposite surface of the glass because they return to their original velocity.

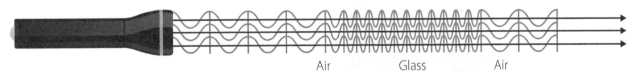
Air — Glass — Air

If the **incident rays** (light rays hitting a surface) hit the surface of the glass at an angle other than 90°, the part of the wave front in the glass is slower than the rest of the wave front. This has the effect of bending the wave front into the glass. Wave fronts resume their original velocity after they leave the glass and bend back to their original direction. This is the cause of refraction in a lens.

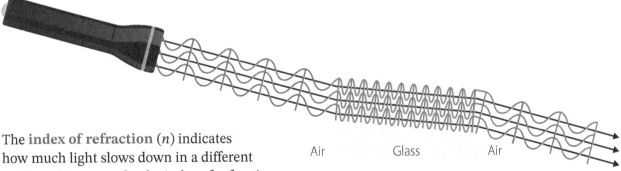
Air — Glass — Air

The **index of refraction** (n) indicates how much light slows down in a different medium. For example, the index of refraction for glass is 1.5. The velocity of light through glass is $v = \frac{c}{n} = \frac{c}{1.5} = 0.667\,c$, which is about $\frac{2}{3}$ the velocity of light where c stands for the velocity of light in a vacuum. This means that the velocity of light in glass is $\frac{300{,}000 \frac{\text{km (kilometers)}}{\text{second}}}{1.5} = 200{,}000 \frac{\text{km}}{\text{second}}$.

Chapter 18 | 145

What if the surface of the lens is not flat but curved? A lens that is curved on both sides is a simple **convex lens**. Convex means to bow outward. These lenses are called positive **converging lenses** because they cause the light rays to come together on the other side of the lens. If separate light rays are perpendicular to an imaginary line down the middle of the lens, they will meet on the other side of the lens at a point called the **focal point**. Notice in this diagram that the light rays are parallel coming into the lens (*Diagram 18.1*).

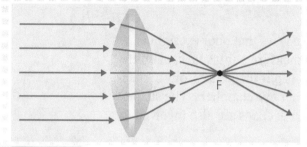
Diagram 18.1

You can indicate the focal point on both sides of the lens because the same thing would happen if light came from the opposite direction. If the object observed is farther from the lens than the focal point, a real image (an image that can be projected onto a screen) will appear on the opposite side of the lens upside down. In this diagram, the light rays coming from the top of the image, a candle flame, are shown passing through the lens and forming a real image on the other side of the lens (*Diagram 18.2*).

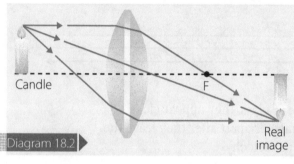
Diagram 18.2

If the object is placed closer to the lens than the focal point, a magnified virtual image (an image that cannot be projected onto a screen) is formed on the same side of the lens as the object. This is done by looking at an object through a magnifying lens. In this diagram, you would be observing from the right side of the lens. The solid lines on this diagram are the light rays. The dashed lines go back to the top of the virtual image, which is what you see when looking from the right side of the lens. It is called a virtual image because you can see it through the lens, but it will not project onto a screen (*Diagram 18.3*).

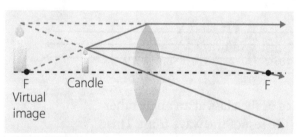
Diagram 18.3

In summary, the simple convex lens curves outward and causes the light rays to converge (come together). If the object is closer to the lens than the lens' focal length, the object is magnified. In your eye, muscles pull on the lens (making it thinner) or relax (allowing it to get wider) in focusing. When you look at objects closer to you, your lens needs to get thinner (lowering the focal length) so that you can focus on them. This is called **accommodation** in physiology. As we get older, the muscles do not respond as well. It helps to have a magnifying lens built into bifocals where the object ends up closer than the focal length of the lens.

Negative **diverging lenses** are concave — they are curved inward on both sides. Light rays hit a surface curved away from them, so they are refracted outward. If dashed lines are drawn back from the light rays, they converge onto a focal point on the side where the light rays entered (*Diagram 18.4*).

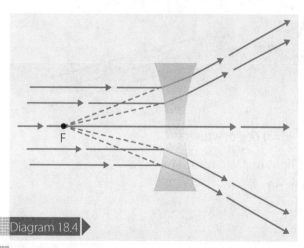
Diagram 18.4

In **far sighted** (**presbyopia**) people, the shape of the eyeball causes it to focus objects behind the retina, so they need to converge the light rays over a shorter distance to the retina. They can see distant objects well, but not near objects. Positive converging lenses are used for this purpose.

Near sighted (**myopia**) people see near objects well, but do not focus well on distant objects. Light rays focus in from of the retina. They need negative diverging lenses to cause the light rays to diverge (spread out). This is done with **concave lenses** (*Diagram 18.5*).

You can bring light rays together with positive converging lenses (convex) and spread them out with negative diverging (concave) lenses.

A microscope is used to view very small objects that are close, and telescopes view objects that are very far away. They both use two converging lenses — the ocular (eyepiece) and the objective lens nearer the object to be viewed. A microscope uses an eyepiece that magnifies the image (both lenses magnify it). A telescope uses an eyepiece that reduces the image rather than magnifying it. The objective lens of a telescope forms a real image of a very distant object and brings it closer to the observer. The eyepiece brings it into focus on the eye's retina. Greater magnification can, however, be achieved by changing the eyepiece of a telescope. The **aperture** is the opening through which light passes. If you have an eyepiece with a smaller aperture, a smaller portion of the image coming through the objective lens is viewed. This is seen as a larger image on the retina giving greater magnification. The downside of it is that there is less light to see with and it is much harder to focus a narrower beam of light onto your retina (*Diagram 18.6*).

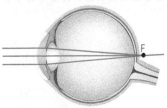

Far sighted focal point behind retina

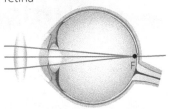

With converging biconvex lens focal point is at the retina

Diagram 18.5

The magnification of a microscope is found by multiplying the magnification of the objective lens (which is marked on the lens) by the magnification of the eyepiece (which is usually 10×). The magnification of a telescope is found by dividing the focal length of the objective lens by the focal length of the eyepiece ($M = \frac{f_o}{f_e}$). The focal lengths are marked on the telescope and the eyepiece. The most critical factor to look for in a microscope or a telescope is the resolution. This is the clarity of the image. Great magnification can just make a fuzzy image into a big fuzzy image.

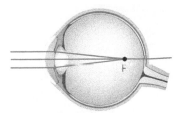

Near sighted focal point in front of retina

A disadvantage of using lenses in telescopes and binoculars is that color aberrations can occur where the edges of the lenses act as prisms and separate some of the colors in light. if lenses are used in very large telescopes, they can be very heavy. These problems are overcome by using mirrors instead of some of the lenses. There are no false colors produced by mirrors, and they can be quite thin and not weigh as much. This will be described in the next chapter.

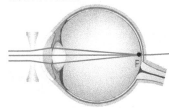

With diverging concave lens focal point is at the retina

Diagram 18.6

Another interesting example of refraction is a **mirage**. This is where there is a layer of hotter air near the ground on a hot day. Hot air is less dense than cooler air, so light travels faster through hotter air. This causes light reflected from an object (like a tree or large rock) to form a virtual image, causing the object to appear as if it is reflected in water even though the ground is dry. This makes it look like there is liquid in the hot, dry desert when there is none. You can see this mirrored-like image at work on the front page of this chapter.

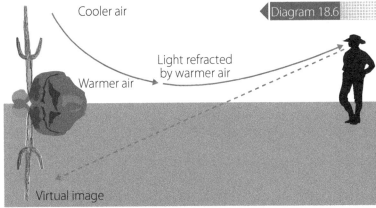

Several years ago, I had a blind student in an astronomy class. We were out at night looking at the moon through a telescope. I was using a moon filter because the moon was so bright. She asked if she could try to look at the moon. I removed the filter, hoping to get a brighter image for her. She looked into the eyepiece and screamed. At first, I wondered what had happened. She saw the moon for the first time in her life. The intensity of the light that reflected from the moon was bright enough to stimulate her retina. She stared with joy. I do not think that there was a dry eye in the class. We need to repent of the sin of taking such blessings for granted. God has blessed us in so many ways.

LABORATORY 18

Lenses — Refraction

REQUIRED MATERIALS

- Laser pointer
- 150 mm focal length double convex lens
- 50 mm focal length double convex lens
- 150 mm focal length double concave lens
- Metric ruler

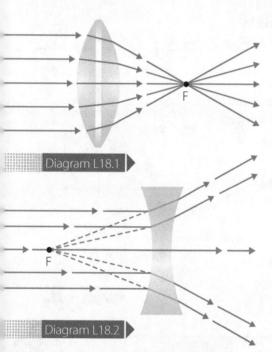

Diagram L18.1

Diagram L18.2

Introduction

A double convex (bows outward in the middle) lens has a focal point on the opposite end of the lens (*Diagram L18.1*).

A double concave (bows inward in the middle) lens has a focal point on the same side of the lens as the light source (*Diagram L18.2*).

A telescope and microscope consist of two convex lenses. One is the objective lens closest to the object and the other is the ocular (eyepiece). The lenses are of different focal lengths, which determines their magnification ability. The difference between a telescope and a microscope is whether or not the eyepiece is the lens with the lower focal length.

Purpose

This exercise demonstrates the focal lengths of double convex lens of differing diameters. The focal length of a double concave lens is demonstrated and contrasted to that of a double convex lens. The arrangements of lenses in telescopes and microscopes is determined.

Procedure

observe

1. Take the 150 mm (millimeter) focal length double convex lens and hold it 15 cm (centimeters which is 150 mm) from a wall. With the laser pointer, point it at the lens close to the outer edge in a circle around the lens. Be sure not to allow the laser to point near anyone's face. A laser can blind you!

   The laser should be focused in a dot in the middle as it goes around the lens because it should focus in a dot in the middle at the focal point. You may need to bring the lens a little closer or farther from the wall.

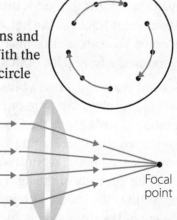

Focal point

148 | Physics

question

2. How would you describe your observations.

3. Point the laser in toward the middle of the lens. Does it still focus on the same spot on the wall?

research

4. Do the same thing with the 50 mm focal length double convex lens. This time see if you can find the focal length by moving the lens back and forth from the wall. The laser should point to the same spot on the wall when the lens is at the focal length distance from the wall. Point the laser around the lens like before. When you get the laser to point to one spot on the wall, measure the distance from the lens to the wall — 50 mm is not very far. It is 5 cm.

5. Hold the 150 mm focal length double concave lens 10 cm from the wall. Point the laser around the outer edge of the lens and observe the pattern that it makes on the wall. Point the laser at different points on the lens. Describe what the concave lens does to the laser beam. Try moving the lens closer and farther from the wall. What happens?

6. Look at an object in the room at least a meter away through the 150 mm focal length double convex lens. Move the lens back and forth until you get the object in focus. Measure the distance from the lens to the object in cm. You may need to get closer to the object. What is the relationship between the distance to the object in focus and the focal length of the lens?

7. Look at a distant object outside. Look through both convex lenses at the object. Try it with the thicker lens closer to you and then try it with the thinner lens closer to you. Move the lenses back and forth as you try to focus on the distant object. Be patient because this may take a bit of moving the lenses back and forth. Which combination of lenses and distance between the lenses worked best to bring the object into focus?

hypothesis

8. This arrangement of lenses is how you would set up a refracting telescope.

experiment

9. Follow the same procedure as in step 5, except this time look at a very small object very close to you.

analyze

10. Which arrangement of lenses and distance between the lenses worked best? How far away was the lens that was closest to the object? This arrangement represents how you would use the lenses to make a microscope. Write a sentence summarizing the difference between the lenses of a telescope and a microscope.

11. Try looking at an object in the room through the concave lens. What happens?

conclusion

12. Try to explain your results.

Lens 1

Lens 2

Laboratory 18 | 149

CHAPTER NINETEEN

DISPERSION OF LIGHT AND REFLECTION

OBJECTIVES

At the conclusion of this lesson the student should have an understanding of

- Index of refraction
- Rainbows
- Diffraction of light
- Iridescence
- The law of reflection
- Flat mirrors
- Concave mirrors
- Convex mirrors
- Critical angle
- Telescopes
- Radio telescopes

Things are not always what they appear to be. It may seem like the sunlight is bright and hot and all you want is to find some cool shade. But within the bright light of the sun's rays is the beauty of a rainbow.

> ... while we do not look at the things which are seen, but at the things which are not seen. For the things which are seen are temporary, but the things which are not seen are eternal (2 Corinthians 4:18).

You will be told many times in life that only what can be seen and tested with repeatable experiments is real. You cannot see God, so you will be told that He is not real. But, when you are tempted to believe it, think of the sunlight and its hidden rainbow that is only revealed in the moisture in the air. God used a rainbow as His covenant with Noah that He would not flood the whole earth again.

The refraction of light, as discussed in the last chapter, occurs as light slows down while passing through another medium. The velocity of light in a medium other than air depends on light's frequency. Each of the colors of a rainbow has its own frequency. Violet travels through transparent materials about 1% more slowly than red light. This is because violet light interacts with the electrons in the atoms of the transparent material more than red light. It is like you stopping to talk to everyone along the way when you are trying to get somewhere. The other colors, blue, green, orange, and yellow are between violet and red. Many of us do not see violet very well, so it appears that the spectrum is from blue to red.

When light passes through a prism, it is refracted twice: when it enters and when it leaves the prism. This gives a greater separation of colors, making them easier to see.

The index of refraction is greater for violet and blue than it is for red, so they travel more slowly than red through glass or other transparent materials. Because blue light travels more slowly through glass or plastic, it is bent (refracted) more than red. The differences in the indices of refraction result in the separation, or dispersion, of the colors within white light.

This also accounts for the colors in a rainbow. Raindrops act like prisms, dispersing light. Light rays are refracted when they enter a raindrop, reflected off the other side of the raindrop, and refracted again when they leave. This results in violet rays being bent at 40° when they leave a raindrop and red rays being bent at 42° when they leave a raindrop. When you see violet and blue in a rainbow you do not see red from the same raindrop because it focuses below you. You see the red from raindrops above the ones that give you the violet or blue, so the rainbow appears to have blue (maybe you will see violet) on the bottom and red on top, with the other colors in between (*Diagram 19.1*).

Rainbows have a bow shape because the 40° and 42° angles are from the ground to the rainbow. If you trace out from the top of the rainbow to the sides, you get a bow.

Sometimes you see another dimmer rainbow above the first one. This secondary bow forms from double reflection within raindrops. The colors are in reverse order and dimmer because more of the light is absorbed by the additional reflection.

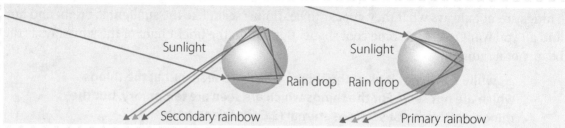

Have you ever seen light go around a corner? Light is not only bent by reflection and refraction, but also by **diffraction**. In a sense, it is going around a corner. Water waves go through a narrow opening and then spread out in the shape of a fan. Light waves do the same thing when they pass through a narrow opening. Radio waves have very long wavelengths, so they readily go around obstructions. This is less apparent with shorter wavelengths of visible light. If light passes through two or more small openings, there are interference patterns between the wave fronts produced which gives different colors.

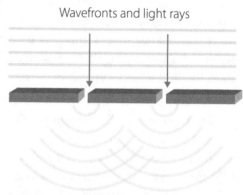

If light is reflected from a surface covered with a thin transparent material, a reflective layer, light reflecting off the two layers will show interference patterns where some of the wavelengths of light are canceled out, leaving the wavelengths of different colors. This is called **iridescence**. You can see this when looking at some fish scales and bird wings.

If the size of an object is as small or smaller than the wavelength of light used, it cannot be seen with a microscope. This is where the shorter wavelength of electron beams is very useful.

You have probably heard of polarized sunglasses. You may even own a pair. Light is emitted in all planes. An illustration, even though it is silly, would be to have ocean waves that are perpendicular to the horizontal and every other plane. Some crystals can filter out the planes of light except one giving polarized light. This greatly reduces the intensity of light. If another sheet of polarized film is placed in front of and perpendicular to polarized light, the light can be eliminated. Then, when the sheet of polarized film is rotated 90°, the polarized light is again visible.

When a beam of light is focused onto a flat mirror, the direction of the reflected light rays is given by the **Law of Reflection**. The angle of the incident light ray (hitting the surface) to the **normal** is the same as the angle of reflection where the normal is an imaginary line perpendicular to the surface of the mirror.

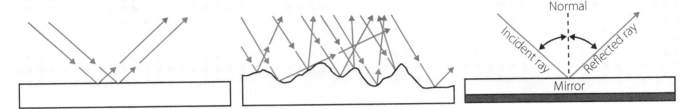

If the reflective surface is not smooth, a **diffuse reflection** is produced.

Concave mirrors produce magnification. When light rays are parallel to a concave spherical mirror, they reflect to a focal point (*Diagram 19.2*).

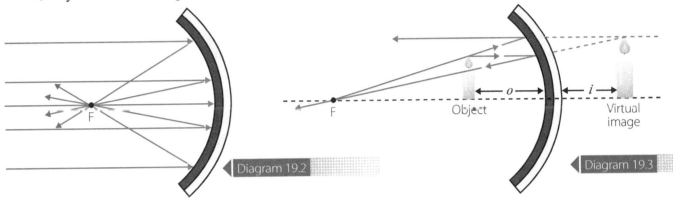

Just like when you look in a flat mirror, if you look in front of a concave mirror closer than the focal point, a virtual magnified image is formed that appears to be behind the mirror (*Diagram 19.3*).

To find the magnification of a concave mirror, you need the distance to the virtual image (i). You cannot measure this directly unless you can climb into the mirror. It is found indirectly using this relationship: $\frac{1}{i} = \frac{1}{f} - \frac{1}{o}$. o is the distance from the object to the mirror and f is the focal length.

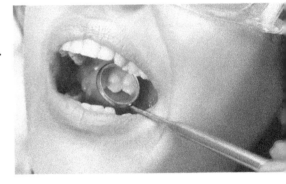

If o is 4 cm and f is 8 cm,

$$\frac{1}{o} = \frac{1}{4} \text{ and } \frac{1}{f} = \frac{1}{8}$$

$$\frac{1}{f} - \frac{1}{o} = \frac{1}{8} - \frac{1}{4} = \frac{1}{8} - \frac{2}{8} = -\frac{1}{8}$$

$$\frac{1}{i} = -\frac{1}{8}$$

$$i = -8 \text{ cm}$$

(it is negative because it goes into the mirror.)

The magnification of the image is $-\frac{i}{o} = -\frac{-8}{4} = 2$. The image is magnified 2×.

If parallel light rays hit a **convex mirror**, they diverge away from the mirror and form a focal point that appears inside the mirror (*Diagram 19.4*).

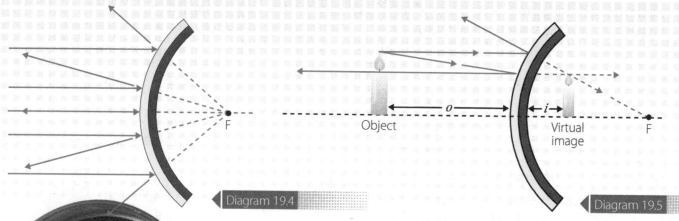

◀ Diagram 19.4

◀ Diagram 19.5

An object placed in front of a convex mirror forms a virtual reduced image that appears inside the mirror (*Diagram 19.5*). These mirrors are not very practical for personal use but are useful for watching for shoplifters in stores because they give a wide view. They are also used to watch for oncoming traffic when going around a corner.

Reflection can also occur from surfaces that do not give images. Whenever you see something, light rays that reflect from the surface of the object pass through the pupil of your eye and interact with the neurons (rods and cones) at the back of your retina. A red object appears red because it reflects red light back to you and absorbs the other colors. Again, this involves the interactions of the light rays with the electrons in the object. What would happen if you tried to see a red object but the color shining on the object had other colors but no red? There would be no red to reflect to you. Would the object be invisible? You know that would not be the case. What happens is that there are electrons in the atoms of objects that are at different energy levels, so they do not reflect just red. They may reflect predominantly red, but there is enough of other colors to see the object — even though it may appear to have a weird dark color.

A different but very useful application of reflection is in fiber optics. When light rays hit a transparent surface directly, they pass through the surface. But if they are at a great enough angle to the surface, called the **critical angle**, they are reflected into the medium. This is useful for laparoscopic surgery to be used like a flashlight in the body. It is also useful for transmitting radio waves over great distances. Remember that radio waves are also electromagnetic radiation. This has widespread use in cable communications.

Mirrors have wide applications in telescopes with distinct advantages over lenses. They can be very thin, so they can be very large and not as heavy as lenses. They do not have color aberrations (distortions) from light passing through the greater curvatures around the edges of lenses. The largest optical telescopes are the two 33-feet diameter reflecting telescopes in the W.M. Keck Observatory in Mauna Kea, Hawaii.

Reflecting telescopes use concave mirrors, flat mirrors, and lens as eyepieces. Light rays enter and strike concave mirrors, reflect to mirrors on the sides of the opening, and reflect forward to a flat mirror that sends the light to the eyepiece. This arrangement of mirrors gives the benefit of a longer focal length while preserving a short barrel.

Radio telescopes are important applications of reflection. Radio waves are electromagnetic radiation but with much lower frequencies and longer wavelengths. Visible light interacts with electrons in transparent and reflective media, and radio waves interact with the electric and magnetic fields of receiving antennae. Radio wave signals are detected from distant space craft with very large radio telescopes. By the time radio waves reach us from deep space, they have spread out, requiring very large antennae (radio telescopes) to pick up enough energy from them. An example is the VLA (Very Large Array) in New Mexico, which consists of 27 radio dishes, each 82 feet in diameter. By taking the stronger signals from interference patterns (interferometry) they can electronically combine them into one huge radio telescope.

There are other exciting applications in astronomy using infrared, ultraviolet, x-ray, and gamma ray telescopes that are constantly expanding. Much is yet to be learned about this great expanse of our universe that God spoke into being.

LABORATORY 19

Dispersion Of Light And Reflection

REQUIRED MATERIALS

- Protractor
- Metric ruler
- Flat mirror
- Penlight
- Red laser pointer
- Prism
- Diffraction grating
- Mirror, concave, 200 mm focal length
- Mirror, convex, 200 mm focal length
- Color paddles

Introduction

A prism uses refraction to separate the different colors of light from each other, based upon their frequencies. The diffraction does the same thing, using diffraction instead of refraction.

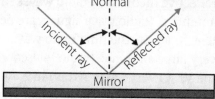

The Law of Reflection states that the angle of incidence between the incident ray and the normal equals the angle of reflection between the reflected ray and the normal.

The magnification of an image in a concave mirror is $M = -\frac{i}{o}$ where i is the distance from the mirror to the object as it appears in the mirror and o is the distance from the object to the mirror that you can measure with a ruler. It is difficult to go into a mirror and measure the distance from the virtual image to the mirror. To do this, use $\frac{1}{i} = \frac{1}{f} - \frac{1}{o}$ as shown in the above example. Then you can use the value of i to calculate the magnification. Just follow the steps in the example.

A convex mirror is a diverging mirror. It can give some rather interesting images. How would you like to have to prepare for a very important meeting in front of a mirror that is part concave and part convex?

Mixing colored lights can give a great deal of variety for a lot of different uses. Mixing paints is a very different phenomenon. If you have a chance, try mixing some of the same color combinations of paints that you used for colored lights and see the differences.

Purpose

This exercise provides experience using light, colors, and mirrors in several different ways. It should help clarify some of the principles of light dispersion and reflection that may otherwise seem rather abstract.

Procedure

 observe

1. Shine the penlight at different angles at the prism. Hold a piece of white paper to use as a screen so that the colors leaving the prism can be seen. Describe the best angle to shine white light on the prism to get a good color spectrum. Shine the penlight through the red lens on the color wheel through the prism and describe what happens. Do it with blue, green, and yellow. Try it with the red laser pointer and describe what happens.

 question

2. What conclusions can you make from these observations?

 research

3. Shine the penlight through the diffraction grating. Again, hold a piece of white paper to act as a screen. Describe what you see. Shine the penlight through the red lens on the color wheel through the diffraction grating and describe what happens. Do it with blue, green, and yellow. Try it with the red laser and describe what happens.

 hypothesis

4. What conclusions can you make from these observations?

 experiment

5. Shine the red laser at an angle at a flat mirror. Place a ruler perpendicular to the mirror where the laser hits the mirror to act as the normal. With a protractor, measure the angles of incidence and reflection. Do your observations obey the Law of Reflection? Explain.

6. Hold an object near a concave mirror where you get a clear magnified image. The focal length of the mirror is 200 mm or 20 cm. Measure how far the object is from the mirror. From this information, determine the magnification of the object. Does your result seem reasonable from your observation? If you come up with a magnification of 2, does it look like it is magnified by 2?

analyze

7. Hold the object near the convex mirror. What does it look like in the mirror? Would you use a convex mirror to make an image of you to photograph?

conclusion

8. Play around with combining different combinations of the color wheels to see what colors you can make while shining white light from the penlight through them.

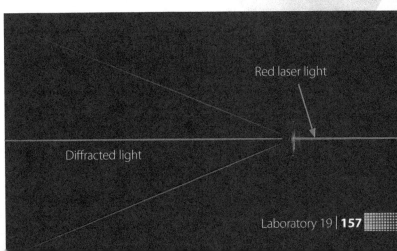

CHAPTER TWENTY

ELECTRIC CIRCUITS 1

OBJECTIVES

At the conclusion of this lesson the student should have an understanding of

- Electric circuits
- Current, voltage, and resistance
- Open and closed circuits
- Circuit diagrams
- Direct and alternating currents
- Circuits in series
- Circuits in parallel
- Voltmeters and ammeters
- Electric power
- kWh

Do you remember the last time you rubbed your shoes across a carpet on a dry, warm summer day and got a shock when you touched a doorknob or sleeping cat? That was from a build up of negative charged electrons on your body. It is called static electricity — it accumulated on your body but was not flowing through a conductor, like a copper wire.

An electric **current** is the flow of electric charge. It can be the flow of positive charges like the movement of positive charged sodium ions (Na+) in a solution or the flow of negative charges like the flow of electrons in a wire. The path that electrons take in a wire is called a **circuit**.

The direction of current flow is indicated as the direction of the movement of positive charges. When electron flow is involved, the current is said to flow in the direction opposite to that of negative charges, even though the overall movement of electrons is in the opposite direction. This allows the same equations to be used in all situations (when dealing with the movement of positive ions or negative electrons). The standard symbol for current is the letter I (*Diagram 20.1*).

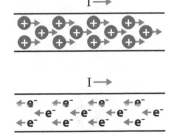

Diagram 20.1

Water flows down a river, rather than uphill, because of gravity. When a book is lifted and placed on a shelf at height h, its gravitational potential energy is mgh in units of joules. When the book falls, its potential energy becomes kinetic energy ($\frac{1}{2}mv^2$) until it hits the floor. Electric charges do not move because of gravity, but rather because of the attraction of opposite charges and the repulsion of like charges. Electric potential is measured in units of **volts**, which is joules per coulomb (a coulomb is the charge of 6.25×10^{18} electrons). Current is measures in units of amperes or amps, as it is usually called. One ampere is $1\ \frac{\text{coulomb}}{\text{second}}$.

A difference between gravity and the attraction of charges is that gravity is in one direction, but electric potential can go either direction depending upon the charges. Charges move between two poles (positive and negative). The positive pole, like on a battery, is the source of positive charges and the negative pole attracts positive charges. Therefore, you must have a complete or **closed circuit**. Consider this circuit with a flashlight battery, a switch, and a bulb.

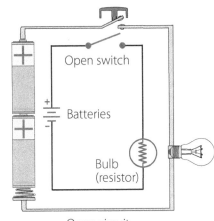

Open circuit

The symbols used in the following diagram are standard symbols. When the switch is down connecting everything together, the circuit is closed. If the switch is up so that the wires are not connected, it is an **open circuit** (the switch is open) and the wires are not all connected — the current cannot get from one pole on the battery to the other. When the circuit is closed and everything is connected, electrons move from the negative pole of the battery to the positive pole of the battery and the current is said to flow from the positive pole to the negative pole.

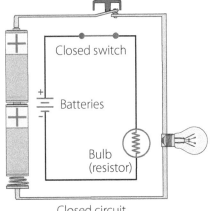

Closed circuit

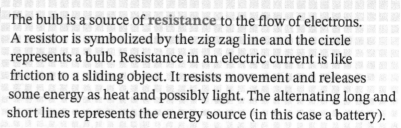

The bulb is a source of **resistance** to the flow of electrons. A resistor is symbolized by the zig zag line and the circle represents a bulb. Resistance in an electric current is like friction to a sliding object. It resists movement and releases some energy as heat and possibly light. The alternating long and short lines represents the energy source (in this case a battery).

Have you wondered how the lights come on almost instantaneously when turn on a light switch? The electrons that leave the negative pole of the battery do not end up on the positive pole. They do not move through a wire like water flowing through a pipe. Metal atoms, which are usually good conductors of electricity, have a few of their outer electrons moving about between the atoms. Consider the analogy of a pipe packed with marbles single file from one end to the other. When you poke a marble in one end, one falls out the other end. When an electron leaves the negative pole of the battery and goes into the wire, one leaves the other end of the wire going onto the positive pole of the battery. Because some of the electrons in a metal wire are mobile, they can do this. Even so, they do not take a direct route but zig zag around a lot between the metal atoms.

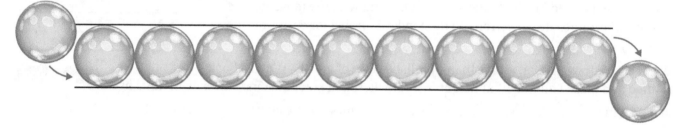

In the circuit diagrammed on the previous page, most of the energy loss is at the bulb, which is why it is called a resistor. The wires and batteries do offer some resistance, but it is very small compared to the bulb. Longer wires have more resistance than shorter wires because more electrons must move through the length of the wire. Also, thicker wires offer less resistance that thinner wires because the electrons have more options to move around metal atoms than in thinner wires.

Electrical resistance is measured in units of **ohms** (symbolized by the Greek letter omega Ω). The opposite of resistance is conductance which is measured in units of mhos (who would have guessed?): $\frac{1}{ohm}$ = mho.

In 1826, Georg Simon Ohm discovered the relationship that came to be known as **Ohm's Law**.

$$\text{Current} = \frac{\text{Voltage}}{\text{Resistance}} = \frac{\text{Volts}}{\text{Ohms}}$$

$$I = \frac{V}{R}$$

When the circuit is closed and current flows through the bulb, the bulb glows and gets hot. Producing light and heat takes energy. It is said that there is a **voltage drop** over the bulb. The amount of voltage lost as the current flow through the bulb is . . .

$$\Delta V = IR$$

The symbol Δ means "a change in" so ΔV means a change in voltage or the drop in voltage.

A battery supplies a **direct current** (DC) because the current flows in only one direction. Household currents are **alternating currents** (AC) because they go back and forth.

In some circuits, when more than one battery or resistor is used, they can be connected in **series**. If you use two batteries and two bulbs, the circuit looks like this (*Diagram 20.2*).

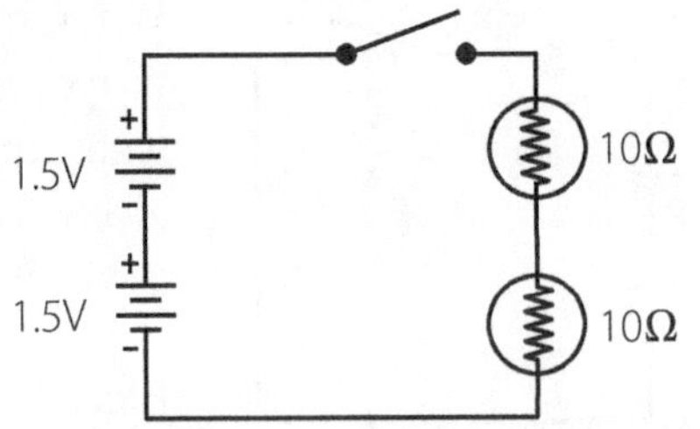

The total voltage is the sum of the two batteries or 3V. if each bulb has a resistance of 10Ω, the total resistance is 20Ω. What is the current of this circuit?

$$I = \frac{V}{R} = \frac{3V}{20Ω} = 0.15 \text{ A}$$

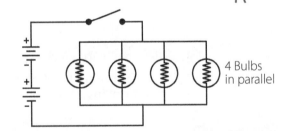

Diagram 20.2

The more bulbs there are in a series, the dimmer each bulb will be because each bulb gets less energy.

If 1 bulb is used, what would be the current available to the 1 bulb?

$$I = \frac{V}{R} = \frac{3V}{10Ω} = 0.30 \text{ A} \text{ — which would produce a brighter bulb.}$$

The voltage drop of the 1 bulb is 3V. When the circuit is complete, the voltage drop equals the voltage provided by the power source. If 2 bulbs are used, the voltage drop across each bulb is $1\frac{1}{2}$ volts.

Another disadvantage of using bulbs in series is that if one bulb burns out, the circuit is open and all the bulbs go out. This is what used to happen with the older design of a series of lights that were hung on a Christmas tree or on the house. There were many bulbs in a series, so the burned-out bulb had to be found and replaced. That could be quite a job.

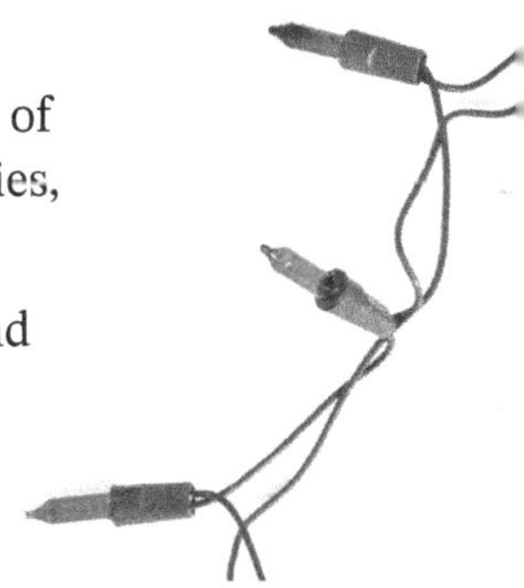

What if 2 batteries and 4 bulbs were used? The total voltage would be 1.5V + 1.5V or 3V and the total resistance would be 40Ω (10Ω + 10Ω + 10Ω + 10Ω). The current would be . . .

$$I = \frac{V}{R} = \frac{3V}{40Ω} = 0.075 \text{ A}$$

Modern Christmas lights are wired in **parallel** instead of in series. This is why if one bulb burns out the others stay lit. If 2 batteries are wired with 4 bulbs in parallel, it would look like . . . (*Diagram 20.3*).

The total voltage is 3V (1.5V + 1.5V) and the total resistance is found by . . .

$$\frac{1}{R_{Total}} = \frac{1}{R_1} + \frac{1}{R_2} + \frac{1}{R_3} + \frac{1}{R_4} = \frac{1}{10Ω} + \frac{1}{10Ω} + \frac{1}{10Ω} + \frac{1}{10Ω} = \frac{4}{10Ω} = 0.4 = \frac{1}{R_{Total}}$$

The total resistance of the four resistors is $\frac{1}{.4} = 2.5$ W.

What is the current of this circuit?

$$I = \frac{V}{R} = \frac{3V}{2.5Ω} = 1.2 \text{ A}$$

There is 1.2 A of current flowing through every bulb since the current has equal access to each bulb.

4 Bulbs in parallel

Diagram 20.3

The voltage and current is measured with a multimeter. First set the dial on the function that you want to measure. To measure the voltage drop across a bulb place the leads of the meter in parallel with the bulb. Be sure that the positive lead from the meter is on the side of the bulb facing the positive pole of the battery. The voltmeter has a large resistance, otherwise the current would flow through the meter instead of the bulb. (*Diagram 20.4*).

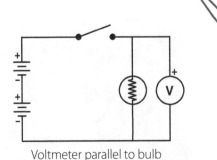

Voltmeter parallel to bulb

Diagram 20.4

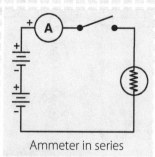

Ammeter in series

Diagram 20.5

When measuring current, switch the dial on the multimeter to amperes in the proper range and it serves as an ammeter. If you wanted to measure the flow of water in a stream, you would place the flow meter in the stream. To measure electric current, place the ammeter meter in series in the circuit between the bulb and the positive pole of the battery (*Diagram 20.5*).

Electric power is $\frac{joules}{second}$ = watts. It is found by multiplying volts × amperes. Volts are $\frac{joules}{coulomb}$ and amperes are $\frac{coulombs}{second}$, so

$$\frac{joules}{coulomb} \times \frac{coulombs}{second} = \frac{joules}{second}$$

If P is power, $P = \Delta VI$ and by Ohm's Law $\Delta V = IR$, $IR \times I = I^2R$. The power input equals the power output which is another way of saying that the voltage supplied by the battery equals the sum of the voltage drops of the resistors.

What is the power used by 3 10Ω bulbs in series powered by 2 1.5V batteries? The total resistance of the bulbs is 30Ω and the total voltage is 3V.

$$I = \frac{DV}{R} = \frac{3V}{30W} = 0.10 \text{ A}$$

The power $P = I^2R = 0.10A^2 30Ω = 0.3$ watts.

What if you had 3 10Ω bulbs in parallel with 2 $1\frac{1}{2}$ batteries? What would the power usage be?

$$\frac{1}{R_{Total}} = \frac{1}{10Ω} + \frac{1}{10Ω} + \frac{1}{10Ω} = \frac{3}{10Ω} \text{ and } R = \frac{10}{3Ω} = 3.3Ω$$

$$I = \frac{DV}{R} = \frac{3V}{3.3W} = 0.9 \text{ A}$$

$$P = I^2R = 0.9A^2 3.3Ω = 2.67 \text{ watts}$$

Utilities charge for electric power usage by the kWh or **kilowatt hour**. Light bulbs are labeled according to their power usage. If you had a 150-watt bulb on for 12 hours, how many kWh would that be?

$$(150 \text{ watts})(12 \text{ hours}) = 1800 \text{ watt hr} = 1.8 \text{ kWh}$$

If the power company charged 10 cents for each kWh, that would cost $1.8 \times 10 = 18$ cents. Usually, power companies have a staggered cost scale where a lower rate is paid up to a number of kWh and more for the kWh above that. Some companies have several layers in the fee structure.

What would it cost to run an electric clothes dryer for 4 hours that used 5400 watts of power, on a 220V circuit that would use 25A?

$$I = \frac{DV}{R} \text{ and } R = \frac{DV}{I} = \frac{220V}{25A} = 8.8\Omega$$

$$P = I^2R = (25A)^2(8.8W) = 5{,}500 \text{ watts} = 5.5 \text{ kW}$$

$$(5.5 \text{ kW})(4 \text{ hr}) = 22 \text{ kWh}$$

$$(22 \text{ kWh})(4 \text{ hr}) = 88 \text{ kWh}$$

$$88 \text{ kWh} \times 10 \frac{\text{cents}}{\text{kWh}} = 880 \text{ cents}$$

or

8 dollars and 80 cents

This is more than the cost for using the light bulb.

What if you had the dryer on a 110V circuit instead?

$$I = \frac{110V}{25A} = 4.4\Omega$$

$$P = (25A)^2(4.4W) = 2{,}750 \text{ watts} = 2.75 \text{ kW}$$

Using about half the power, it would probably take twice as long to dry the clothes. Power is the rate that energy is being used, so if half the power is being used, it would take twice as long to use the same amount of energy.

Electricity can be a blessing and a curse. It can power machines and convey information rapidly over great distances. It is an example of how exact and efficient God's creation is in spite of the fall. It does not take much energy for an electric current to travel large distances. Think of how fast and efficiently information is carried over the internet. On the other hand, generating electricity can harm the environment and has also been the cause of several destructive fires.

Just where do we get our power to switch on lights or to dry out our clothes in the dryer? One source is hydroelectric power, generated from water running through a dam. The Diablo Dam is located in Whatcom Country, Washington, and provides electricity to Seattle. It produced its first electricity back in 1936, and is operating still. The water rushing through the dam operates two main generators, each with a capacity of just over 64 megawatts. One megawatt is equal to one million watts!

LABORATORY 20

Electric Circuits

REQUIRED MATERIALS

- Insulated copper wire
- Switch
- 3.7 V bulbs (2)
- Bulb holder (2)
- AA battery holder
- AA batteries (2)
- Multimeter

Introduction

This lab exercise involves 2 AA batteries and one or two bulbs in a circuit which is controlled by a switch. When the switch is open, no current can flow through the circuit because there is a break in the connections. When the switch is closed, the current can flow through the circuit. The voltage is measured from one end of a bulb to the other. This is a **voltage drop** because some of the electrical potential energy is used by the bulb. It is like a book falling part way to the floor using part of its potential energy.

When measuring the voltage drop, a voltmeter is connected in parallel to the bulb because the difference in voltage between the two points is being measured. When measuring the current, the ammeter is placed in line with the current so that the current flows through it the same way that you would place a flow meter in a stream to measure the rate of water flow.

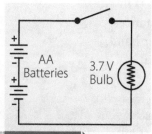

Diagram 20.1

If you are curious, you can measure the voltage across the batteries. If you measure it when the switch is open, you should get 3V ($1\frac{1}{2}$V + $1\frac{1}{2}$V). This is a property of the chemical reactions in the batteries. If you measure the voltage across the batteries when the switch is closed, it should read a bit less than 3V (maybe around $2\frac{1}{2}$V) because as the current flows through the batteries there is some internal resistance.

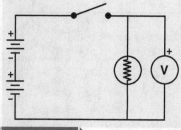

Diagram 20.2

In any electrical circuit, the current flow is indicated from the positive pole of a battery to a negative pole of a battery. The flow of electrons, however, is in the opposite direction — from the negative pole (where there is an excess of negative electrons) of the battery to the positive pole.

Purpose

This lab exercise is to provide experience setting up an electrical circuit with bulbs in series and parallel and measuring their voltage drops and current.

Procedure

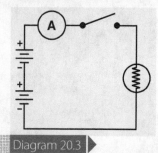

Diagram 20.3

observe

1. Using lengths of the insulated copper wire (at least 1 to 2 inches long per section), connect the battery holder with 2 AA batteries, the switch, and a bulb holder with a 3.7V bulb. Connect the wires to the screws on each of the different parts (*Diagram L20.1*).

Test your circuit by closing the switch to be sure that the bulb shines. Then open the switch so that you do not drain the batteries.

For all of these procedures, attach the red wire of the multimeter to the outlet on the meter marked VΩmA, and the black wire to the outlet on the meter marked COM. When testing the circuit, the red lead from the meter is closest to the positive poles of the batteries.

? question

2. As you work through this lab, describe what you are doing and your results to your teacher. Record your results in your report. Describe the direction of current flow and the direction of electron flow.

3. Disconnect the wire connected to the bulb holder and connect the red lead to the wire going to the battery holder and the black lead to the bulb holder. Set the dial on the meter to 200mA (the maximum reading on the meter is 200 mA or 0.2 A). If the reading on the meter exceeds 200 mA, set the dial to 10 A. This means that the maximum reading is 10 A. Close the switch and read the current from the meter. Record this value. Open the switch (Diagram L20.2).

research/hypothesis

4. Remove the multimeter from the circuit and restore the wire connections as before. Turn the dial on the multimeter to DCV 20 V. Place the red lead of the multimeter on the side of the bulb holder close to the positive poles of the batteries and the black lead on the opposite side of the bulb holder. Close the switch so that the bulb glows and read the voltage from the multimeter. Record this value. Open the switch (Diagram L20.3).

experiment

5. Add a second bulb holder with a bulb into the circuit in series with the first bulb as shown in this diagram (Diagram L20.4).

6. Repeat step 2 with 2 bulbs in series instead of 1 bulb. Record the value of the current with two bulbs (Diagram L20.5).

7. Repeat step 3 with 2 bulbs in series instead of 1 bulb. Measure the voltage drop of each bulb separately and together (Diagram L20.6).

8. Rewire the circuit so that you have 2 bulbs in parallel instead of series (Diagram L20.7).

9. Repeat step 2 with 2 bulbs in parallel instead of 1 bulb. Record the value of the current with two bulbs connected in parallel (Diagram L20.8).

10. Repeat step 3 with 2 bulbs in parallel instead of 1 bulb. Measure the voltage drop of each bulb separately (Diagram L20.9).

analyze

11. From the current and voltage of the circuit in step 2, calculate the resistance of the bulb. Use Ohm's Law $R = \frac{V}{I}$. Show your work in your report.

12. From your results in this lab exercise, describe the meaning of the terms *current*, *voltage*, and *resistance*. How did placing the bulbs in series affect the current?

conclusion

13. How did placing the bulbs in parallel affect the current?

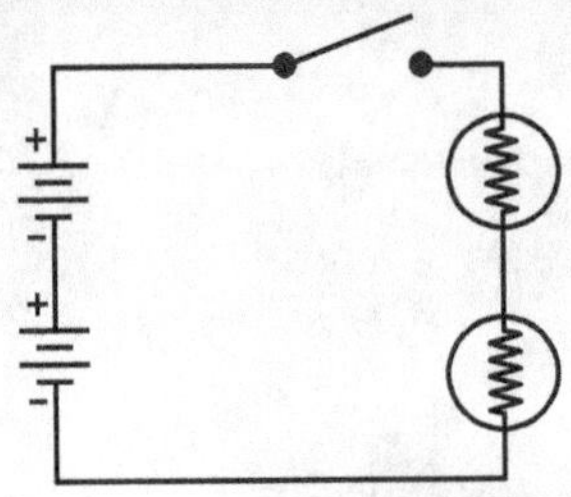

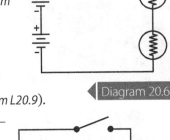

Diagram 20.4

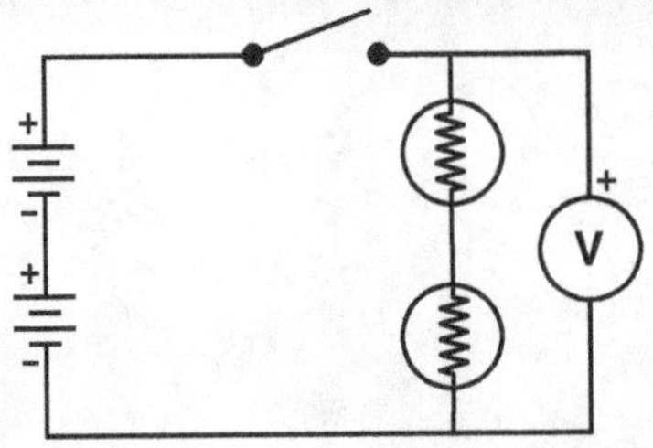

Diagram 20.5

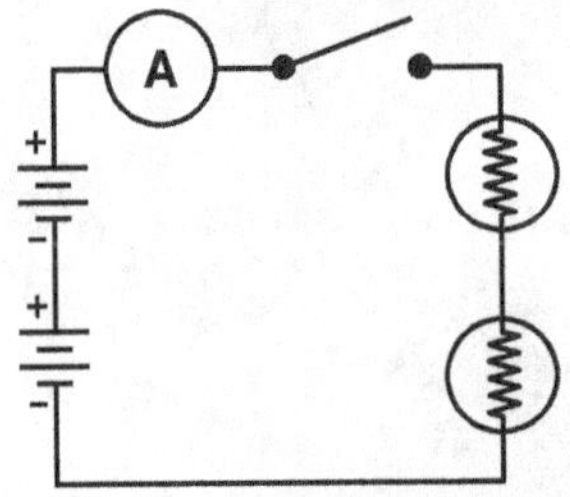

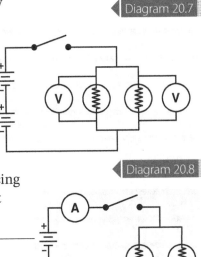

Diagram 20.6

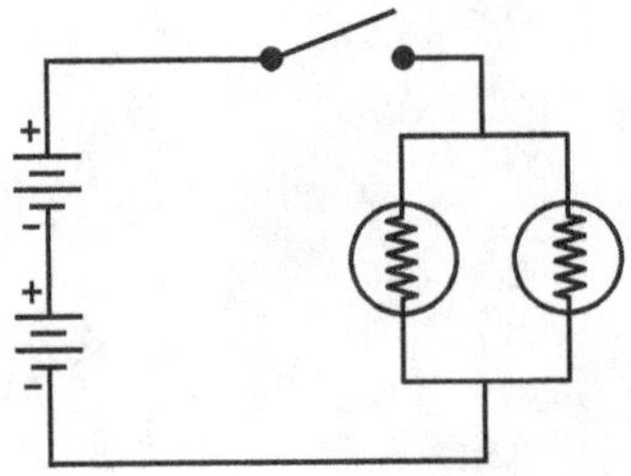

Diagram 20.7

Diagram 20.8

Diagram 20.9

CHAPTER TWENTY-ONE
ELECTRIC CIRCUITS 2

OBJECTIVES

At the conclusion of this lesson the student should have an understanding of

- Alternating currents
- Electric shock
- Lightning
- Electric power
- Capacitors
- Integrated circuits
- Semiconductors
- Rectifiers
- Transistors
- Superconductors

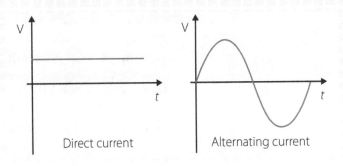

Direct current   Alternating current

An **alternating current** (AC) flows or alternates back and forth. In the United States, the current alternates back and forth about 60 times per second. This is called the frequency as 60 Hertz (Hz). The voltage peaks out at 160 volts and then diminishes down to zero and goes the other way up to 160 volts and back down again. The average (or effective) voltage in each direction is 110V. This is called the root mean square (RMS) value.

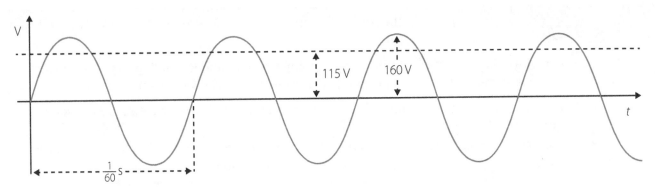

This curve is called a sine wave after the trigonometry function. Household circuits have alternating currents and are always connected in parallel. There are several separate circuits in a home where several outlets are connected to the same circuit. As more appliances are plugged into the same circuit, the total current flowing through the circuit increases. It is better to have more circuits with fewer devises attached to each. When the current gets too high in one circuit, the wires get hot and will burn off the insulation around the wires and start a fire. Each circuit has a **circuit breaker** (fuses are also used) that cuts off the flow of current through the circuit if the amount of current gets too high. Circuit breakers can be rated at 10 A, 15 A, or 20 A. Sometimes a 30 A breaker will be used. The circuit breakers are in a circuit panel that is in the garage, basement or outside. What are the advantages of having the appliances on the same circuit in parallel instead of in series?

AC circuits usually have higher currents than DC circuits. Batteries with multiple cells of chemical reactions can generate higher currents. A car battery (12V) can shock a person, but not a flashlight battery. Electric shock occurs when an appreciable current flows over or through any part of the body. An important principle to remember is that we can survive a high potential (voltage) but not a high current. An electrical current will seek the path of least resistance. Dry outer skin has very high resistance (about 100,000 W), but the fluids within the body offer very little resistance. When the body is wet, the outer resistance is reduced to less than 700 W. This is why a person caught in a lightning storm can survive with multiple burns but still be alive. The resistance outside the body is reduced by water and salts outside the body. This is not a great situation because a person can suffer terrible burns. In a similar manner, people have

Blisters on finger caused by electric shock

been electrocuted by hair dryers and radios falling into bath water. Bathroom circuits today are equipped with circuit breakers that shut off the current when there is a surge in the current flow. Examples of the effects of an electric current on the human body are . . .

Header	Header
0.001 A (1 mA)	Can be felt
0.005 A (5 mA)	Painful
0.010 A (10 mA)	Causes involuntary muscle spasms
0.015 A (15 mA)	Causes loss of muscle control

The wires in household circuits can carry far more current than is lethal. Electricity can be very helpful in everyday life, but if not used wisely can be very dangerous. The cords from appliances have 3 wires. The flat prongs in the end of a cord carry the current back and forth. A third round prong in the cord will carry current to the ground if there is a possibility of shock occurring.

Which would you rather get zapped by — a high current and low voltage or a low current and high voltage? Personally, I would just as soon not get zapped by either.

When electric current is coming from where it was generated, it can travel great distances with a potential of as high as 500K V (500,000 volts) and very low current. Being an alternating current. the actual electrons do not travel very far. A lot of energy (voltage), a short distance (alternating current), and few electrons (low current) makes it very efficient. A number of step-down transformers lowers the voltage and raises the current several times before it enters your home at 110 V. At the circuit box, several separate circuits, each with its own circuit breaker, branch out into the home.

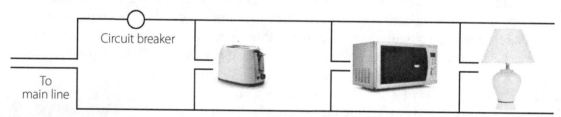

An interesting example of the movement of electric charges involves thunder and lightning. A large fluffy white thundercloud has a build up of positive charge on the top of the cloud and negative charge on the bottom of the cloud. There is a lot of rising of warmer air and falling of cooler air within a thundercloud. This probably causes the charge separation within a cloud. A positive charge is induced on the surface of the earth by the negative charges on the bottom of the cloud repelling electrons farther into the ground. The electric field generated by this charge is several thousand volts per meter. Because the cloud bottoms and the ground are about several hundred meters apart, the total potential difference between the bottom of the cloud and ground is therefore about several million volts! No wonder we get such spectacular lightning storms.

Dry air is a good insulator, but moist air is a good conductor. A flow of charge along a path with the best conductance and shortest distance heats the air and ionizes (removes outer electrons) atoms in the path making the air a much better conductor. This initial flow of charge is called the leader. Several discharges make the air an even better conductor and a very large flow of charge takes place along this path. The light is produced by electrons of the air atoms that jump to higher energy levels and then release this energy as light as they go back down to lower energy levels. The same thing happens in a flame to a lesser extent. The thunderclap is produced by air pushing against air. You will see the lightning before hearing the thunder because light travels much faster than sound. Pointed objects produce stronger electric fields and make good lightning rods. Lightning is as hot as the surface of the sun and produces a ton of nitrates (NO_3^-) that acts as fertilizer from oxygen and nitrogen. A lone or clump of trees can also act as lightning rods. The best place to be in an electrical storm is in a building or automobile or in a ditch or gulley below ground level.

The wattage of an appliance can be used to calculate the current the appliance will use. A color TV may use about 100 W of power.

$$P = I \Delta V \text{ and } I = \frac{P}{\Delta V} = \frac{100W}{110V} = 0.9 \text{ A}$$

The resistance of the color TV can then be found using Ohm's Law as . . .

$$R = \frac{\Delta V}{I} = \frac{110V}{0.9A} = 122 \text{ W}$$

Some examples of wattages (these can vary quite a bit) of some common household appliances are . . .

Appliance	Power (watts)	Current (A)	Appliance	Power (watts)	Current (A)
Stove	6000 W (220 V)	27 A	Coffee maker	1000 W (110 V)	9 A
Clothes dryer	5400 W (220V)	25 A	Toaster	850 W (110 V)	7.7 A
Water heater	4500 W (220 V)	20 A	Light bulb	120 W (110 V)	1.1 A
Electric iron	1100 W (110 V)	10 A	Clock radio	12 W (110 V)	0.1 A

As a rule, heat-producing appliances use more current.

If you have a 20 A circuit breaker on a circuit, would you plug an electric iron, coffee maker, and toaster in that circuit at once? They would pull 10 A, 9 A, and 7.7 A at the same time. That is 26.7 A total and the circuit breaker trips (opens the circuit) and the appliances stop working. This is why it is helpful to have more than 1 circuit in the kitchen.

What if your stove is not on a 220 V circuit, but on a 110 V circuit? The current pulled by the stove would be $\frac{6,000W}{110V} = 55$ A. There are no circuits that high in your home. It would heat up and burn the insulation off the wires and burn the house down. It is better to use the 220 V circuit. They are formed by combining two 110 V circuits.

Computers and other electronic applications make extensive use of timing devices. One of the more common forms is the use of **capacitors**. A capacitor is a rather simple clever arrangement of two conductors separated by insulating material. A power source, like a battery, moves negative charges (electrons) onto one of the conductors and removes electrons from the other conductor, making it more positive. As the charges build up on the conductors, they finally overcome the resistance of the insulating material, and negative electrons jump over from the negative conductor to the positive conductor. The time it takes to build the charge before they jump over is a time delay before the current can flow through the circuit. The word *capacitor* refers to the capacity of the conductors to store electric charge before it is released (*Diagram 21.1*).

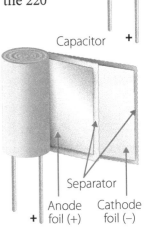

Diagram 21.1

A major breakthrough in electric circuitry involved **integrated circuits** also called IC, chips, and microchips. Basically, this is a means of miniaturizing an enormously large electric circuit into a very tiny space smaller than a postage stamp. This was made possible through the use of a group of elements called **semiconductors**.

Metals are typically good conductors of electric current while non-metals are usually good insulators (poor conductors). Metal atoms have a few electrons that "float" between the atoms and can move through the metal as an electric current. Non-metal atoms attract and keep extra electrons rather than having them available to move about.

There is a small group of elements that are in between metals and non-metals. These are called the metalloids or semiconductors. They include the elements carbon, germanium, and silicon. In their pure form they are not good conductors of electric current. Their resistance is modified by **doping** them with particular impurities. An example is adding a small amount of phosphorus, arsenic, or antimony to silicon. When the phosphorus, arsenic, or antimony atoms bond with the silicon atoms, they have an extra electron that can move about. This makes silicon an **n-type semiconductor**. The letter "n" stands for the negative charges of the extra mobile electrons. When atoms of boron, gallium, or indium are added to a semiconductor it becomes a **p-type semiconductor**. This is because they can acquire an extra electron. The "p" stands for positive because lacking a negative electron is like having an extra positive charge.

Boron, gallium, or indium can move about in the silicon. Because they could acquire another electron, they are called a hole. When these atoms move around, it is said that the "holes move about." This sounds more like something out of a cartoon where someone digs a hole and picks it up and puts it somewhere else. If an n-type semiconductor is placed near a p-type semiconductor, the hole in the p-type semiconductor can move closer to the n-type semiconductor and acquire its extra electron. This produces a small electric current. if they are arranged properly, atomic-sized electric currents can produce a complex circuit. The next question is — how do you place them in the right places? Once the desired pattern is known, it can be replicated many times by covering certain parts and using a technique called photolithography to imprint the desired pattern. There are a lot of details left out here that would come in more advanced courses.

Rectifiers are devices that turn an alternating current into a direct current. A **diode** is a common example of a rectifier. A common semiconductor diode is where n-type and p-type materials are next to each other — when the positive terminal of a power source is connected to the p-type material and the negative terminal of the same power source is connected to the n-type material. Electrons from the power source drive electrons from the n-type material onto the p-type material where the electrons fill the "holes" in the p-type material (the doping atoms acquire the electrons). The doping atoms carry the electrons to the positive side where they go back to the power source and there is a complete circuit. This results in electrons flowing in one direction (direct current) (*Diagram 21.2*).

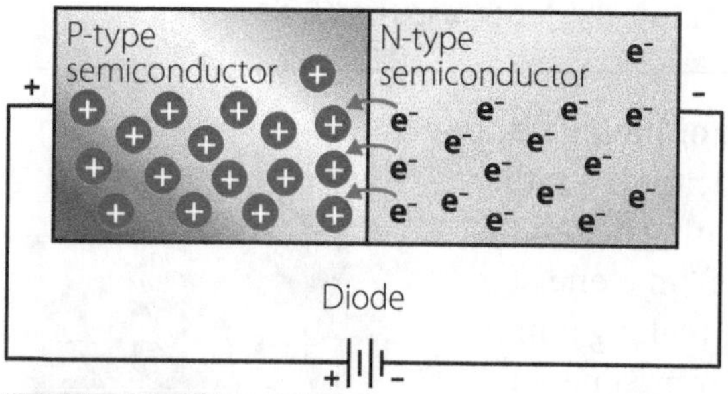

A **transistor** is a combination of two diodes. They amplify or switch electronic signals. It is semiconductor material with at least three terminals that can be connected to an external circuit. Because the output power can be higher than the input power, it is an amplifier. When two diodes are connected by a thin layer of material called a base layer, a small change in the base to the emitter produces a large change in the current from the collector to the emitter (*Diagram 21.3*).

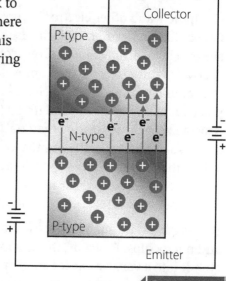

Diagram 21.2

Diagram 21.3

A common application is where a weak signal from a radio antenna is turned into strong signals using semiconductor transistors. These are routinely used in radios and televisions. They are also used as voltage controlled switches (sends a current in a different direction or turns it off) by changing the voltage to the base. There are also field-effect transistors (FET) where a thin layer of n-type material (called a channel) is sandwiched between two layers of p-type material. A current flowing through the n-type channel is controlled by voltage applied to the two p-type channels. These are very common in computer circuitry. When hundreds of transistors and other circuit parts are combined on tiny silicon chips, they form integrated (coordinating together) circuits. All of the knowledge and successful and unsuccessful experiments have gradually, since the late 1800s, contributed to the tools we have today. No one person understands everything that goes into a computer, but the combined efforts of many people have made it possible.

When light strikes a doped semiconductor, it dislodges electrons from the "holes" of n-type material. A solar cell is made with a very thin layer of p-type semiconductor material on top of n-type material. Light penetrates the thin layer of p-type material and dislodges electrons from the underlying n-type material. The electrons then occupy "holes" in the p-type material, producing a weak current. This is called the **photovoltaic effect**. If more current is needed, several solar cells can be linked in parallel or more light can be focused on the cells. If they are connected in series, the voltage will be increased.

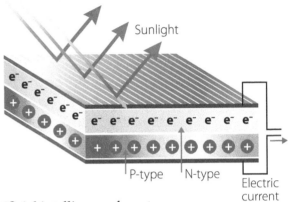

Cross section of a solar cell

Remarkably, all of the efforts in computer technology and artificial intelligence do not come close to what God spoke into existence in the human brain and nervous system. What is even more remarkable and hard to comprehend is that we will retain the ability to think, function, and respond as spiritual beings even after our bodies decay and before we have resurrection bodies.

Superconductors are materials (ceramic materials of metal oxides) that lose all of their electrical resistance below a critical temperature. Some materials have a critical temperature of about 90 K (–183°C). Liquid nitrogen (which is readily available) has a boiling point of 77 K (–196°C) so it can be used to cool these materials to their critical temperature. As superconductors, these materials can conduct electric currents indefinitely without losing energy as heat due to resistance. Some commonly used materials consist of combinations of yttrium (Y), barium (Ba), copper (Cu), and oxygen (O). Another property of a superconductor is the Meisner effect where they exclude magnetic force fields. When a magnet is brought near a superconductor, it is repelled. This enables a small magnet to be levitated above a disc of superconducting material. Weird! Practical applications include producing large electromagnets with tightly wound coils of wire without the tremendous amount of heat they usually generate. These could be used to power and levitate bullet trains and other forms of transportation.

Over time electric circuits were miniaturized in the form of microchips as shown in digital information storage and transmission. This provides more cost effective applications – such as the brightness and longevity of light produced by LED bulbs and the rapid response times of computers in automobiles.

LABORATORY 21

Electric Motors And Solar Cells

REQUIRED MATERIALS
- Electric motor assembly kit
- Solar cell, 1 V, 500 mA
- 3.7 V bulb
- Bulb holder
- Multimeter

Introduction
An electric current supplied to a wire induces a magnetic field. If the wire is surrounded by an external magnet such that the external magnet repels the magnetic field of the wire, the wire can be made to rotate if properly arranged. By assembling the electric motor from the kit and watching it function, you can visualize how the external magnetic field causes the wire to rotate.

A solar cell is made with a very thin layer of n-type semiconductor material on top of p-type material. Light penetrates the thin layer of n-type material and dislodges electrons from the underlying p-type material.

Purpose
This lab exercise demonstrates how the repulsion and attraction of an external magnetic field upon an induced magnetic field in a wire causes the wire to rotate.

An important application of a doped semiconductor is a solar cell. This lab exercise enables the student to measure the photovoltaic effect upon a solar cell.

Procedure

 observe

1. Follow the instructions with the electric motor assembly kit and assemble the electric motor.

 question

2. Describe how you assembled it, demonstrate its use, and explain why it works to your teacher.

 research

3. Connect the leads from the solar cell to the bulb holder from lab 20. Screw a bulb into the holder.

 hypothesis

4. Do you notice any brightness to the bulb from the light in the room shining on the solar cell? Shine a brighter light from a flashlight on the solar cell. Is the bulb very bright? The solar cell is designed to give 1 V of 500 mA (milliamps) or $\frac{1}{2}$ amp of current.

 experiment

5. Connect the multimeter in parallel to the bulb (like in lab 20) and set the dial to DCV 20 V to measure the voltage drop at the bulb, which is the voltage supplied by the solar cell.

 analyze

6. Connect the multimeter in series with the bulb and solar cell (like in lab 20) and set the dial to 10 A to measure the current supplied by the solar cell.

 conclusion

7. Look back into chapter 21 and describe how the solar cell is constructed and how it is producing the electric current.

CHAPTER TWENTY-TWO

ATOMS AND OTHER TINY THINGS

OBJECTIVES

At the conclusion of this lesson the student should have an understanding of

- Scientific models
- Cathode ray tube experiments
- Electrons
- X-rays
- Natural radiation (alpha particles, beta particles, and gamma rays)
- Blackbody radiation
- Bohr's model of the atom
- Quantum mechanics

What can you see that you cannot see? Light. You actually cannot see anything except for the light that reflects from objects and back to your eyes. But what are you seeing? The modern concept of light is that it is a wave and also particles. A wave keeps coming and particles hit a target like snowballs. In an earlier chapter, the topic of scientific models was introduced. We cannot see things as small as individual atoms and electrons. Yet we talk about them as if we know what we are talking about. We cannot see them, but we can see a bunch of them together and we know what they do. A **model** of light is a description of something that would behave like light even though we have no way of knowing if the description is accurate or not. It is like someone describing someone they met for the first time. Even though you have never seen that person, you still get an image in your mind based on the description you hear. If light were a wave, it would do some of the things that waves do — such as wave diffraction and interference. If it were composed of particles instead of waves, it would do the things that we are dealing with in this chapter. This gives us two different models of light and that is okay because a model is not a description of what it is actually like. The model is helpful in that it helps us predict what light would do, even though it does not give us a description of what we cannot see about light. The goal is to predict the behavior of light — not to give a clear description.

Atoms and electrons are also too small to see. Their stories also involve observing phenomena and coming up with models that try to match their behavior. Light, atoms, electrons, and radioactivity are all players in a common drama.

In 1869, Johann Hittorf (1824–1914) reported the results from experiments involving a **cathode ray tube**. A **cathode** is the negative electrode of a power source and an **anode** is the positive electrode. The power source builds electrical potential by driving electrons onto the cathode, making it negative, and removing electrons from the anode, leaving it with positive charges. In 1869, without knowledge of electrons, he observed that when there was high voltage between the cathode and anode in a glass tube with a vacuum, called a cathode ray tube, a glowing beam traveled across the glass tube from the cathode, making a glowing spot seen on the other end of the glass tube. When he brought the north pole of a magnet near the beam, the spot was deflected to the left. The direction the beam moved (to the left) indicated that the beam was composed of moving negative charges (*Diagram 22.1*).

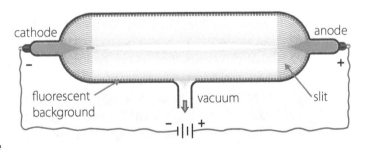

Diagram 22.1

Follow-up studies published in 1897 by the English physicist J.J. Thomson (1856–1940) demonstrated that the particles in the cathode ray beam had a mass $\frac{1}{1,830}$ times that of a hydrogen ion (hydrogen atom without its electron, making it a proton). Today we know the

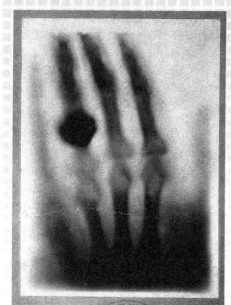

Radiograph of hand with a ring by Wilhelm Roentgen

cathode ray beam to be composed of electrons. Notice that he never saw electrons, but he did see what the electrons were doing.

German physicist Wilhelm Roentgen (1845–1923) noticed in 1895 that when he was working with a cathode ray tube, a piece of fluorescent (which glows when exposed to certain frequencies of light) paper on his workbench glowed when the cathode ray tube was used. The cathode rays were not able to go outside of the glass tube. He covered the tube with black paper and the fluorescent paper still glowed. This meant that the beams in the glass tube were not causing the paper to glow, but something else was being released whenever it was turned on. This effect took place even when the paper was at least 2 meters away from the cathode ray tube. He called this new radiation **x-rays** because x stands for something unknown. He also noticed that he could see the shadow of the bones in his hand when he passed his hand between the cathode ray tube and the fluorescent paper. That must have been weird! This radiation was found to be very high frequency (more energy than ultra-violet) electromagnetic radiation. A few years later this was used to examine broken bones and other tissues in medical applications without realizing the negative impacts of excessive exposure. Some shoe stores used x-rays to let customers see the bones in their feet with the idea of being able to get them a better fit of footwear. It was actually a novelty just to get them into the store.

This was a time of rapid discovery because one thing led to another. In 1896, the French physicist Henri Becquerel, while studying phosphorescent rocks (that glow in the dark after being exposed to ultraviolet light), especially those that had uranium, noticed that photographic film would show dark areas when exposed to these rocks. He realized that these rocks did not have to be exposed to light to expose photographic film and that they could keep doing it indefinitely. He called this radiation **natural radioactivity**.

Marie and Pierre Curie and Ernest Rutherford further pursued this study by boring a narrow hole into a block of lead. They placed a sample of uranium in the hole so that the radiation would come out of the hole in a narrow beam. When the beam was passed through a strong magnetic field, it split into three separate beams. Rutherford named these beams after the first 3 letters of the Greek alphabet — alpha (α), beta (β), and gamma (γ). The α beam (called α particles) were found to be helium (He) atoms minus their two electrons. This made them heavier than the others. The β radiation (called β particles) were high-energy beams of electrons. The γ radiation was very high frequency electromagnetic radiation with more energy than x-rays. This was a huge discovery.

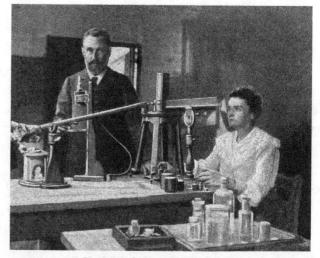

Pierre and Marie Curie in 1904

This led to the later work, discovered in 1911, by Rutherford, who discovered the overall structure of the atom and its nucleus. He bombarded a thin sheet of gold foil with a narrow beam of α particles. Most of the a particles went right through the gold foil; some were deflected at different angles and very few came back in front of the gold foil. This is called a scattering experiment because the α particles were scattered. He concluded that most of the α particles went right through empty space in the gold atoms, and those deflected back toward the front of the gold foil hit the nucleus of an atom where all of the mass was located. The model of the atom developed from this experiment was that atoms were composed of mostly empty space with a massive positively charged nucleus near the center. The proportions were as if the atom were the size of a football stadium, with the nucleus the size of pea-sized rock on the 50-yard line.

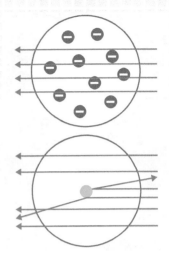

Rutherford's expected results (top) and his observed results (bottom)

With a better understanding of the relationship of the nucleus to the rest of the atom, they wondered about the electrons. About 50 years before Rutherford's work, they had observed the light emitted from hydrogen gas when high voltage was passed through it, making it glow. The light from hydrogen could be separated by a prism or diffraction grating into 4 distinct wavelengths in the visible spectrum — one line in red, one in blue, and two in violet.

In trying to understand more about the electrons, Max Planck (1858–1947) and others studied **blackbody radiation**, where a block of ceramic or metal with a hole in it is heated to high temperatures. The hole itself appeared black but gave off light when heated. At first it became red hot and then white. The wavelengths of light emitted by this object only had certain values. Planck came up with a formula to predict the wavelengths of light emitted that were dependent upon the temperature of the block. These set wavelengths of light were called quanta. Planck suggested that the emitted light had distinct energy values with no energy values in between. The Danish scientist Niels Bohr, who worked with Rutherford, came up with a model to account for the electrons about the nucleus of the atom using these observations. He said that the electrons had set values that he called orbits. In his model, electrons were in set **orbits** in the empty space of an atom around the nucleus. He thought of the orbits as being like planets orbiting around the sun. He tried to predict the energy levels of electrons in other elements beside hydrogen, but these were not always successful even for hydrogen (*Diagram 22.2*).

Max Planck in 1933

Blackbody radiation

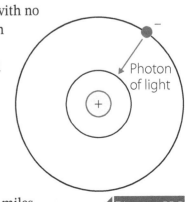

Diagram 22.2

To appreciate the idea of quanta, think of a new car that could only go at certain velocities. The car can go 20 miles an hour, 40 miles an hour, and 60 miles

Chapter 22 | 177

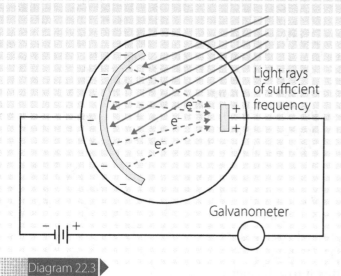

Diagram 22.3

an hour. The car cannot go at the velocities in between. Sounds crazy, doesn't it? You would get a whip lash just going from zero to 20 miles an hour immediately. This represents the energy levels available to an electron. As the electrons went from a higher energy level down to a lower energy level, they gave off the energy difference as light. This is why there were only 4 visible color lines given off by hydrogen. Hydrogen has only one electron, and the colored lines are the energy differences between the possible energies for the electron.

From other studies, De Broglie supported the view that light could be thought of as being particles of energy called **photons**. This idea was given further support later by Albert Einstein. Einstein used the **photoelectric effect** to support the idea that light was composed of photons. Certain metals give off electrons to become part of an electric current when exposed to light. It only worked when the frequency of light was great enough. If light was a wave, the energy from the wave should gradually accumulate and be enough to eject electrons. This was not the case. When light of sufficient frequency (energy) hit the metal, electrons were immediately ejected. It worked well with violet or ultraviolet light but not with lower energy red light. If the light was brighter, more electrons were ejected — more photons with enough energy would be hitting the metal. Higher frequency light caused electrons of greater energy to be ejected (*Diagram 22.3*).

De Broglie further proposed that if light waves can be thought of as particles; then maybe electrons (particles) can be thought of as having wave properties. Remember, these are models. This is not saying that light and electrons are particles or waves, but that they can behave like particles or waves. Particles and waves are approached with different equations when trying to predict their energies and behaviors. This idea of thinking of electrons as waves was picked up by several others, most notably by Erwin Schrodinger who designed a three-dimensional model of electrons in atoms. He treated them as standing waves and designed equations to predict their energy levels and where they were most likely to be found in an atom. He used the term **orbital** to distinguish them from the orbits of Bohr. Like Bohr, he kept the idea that light was given off when an electron lost energy to place it in a lower energy orbital. This is where the light in a flame comes from. But, unlike Bohr, he dealt more with energies of electrons rather than their positions. He incorporated the **Heisenberg Uncertainty Principle**, which states that you cannot know both the velocity and position of an electron at the same time. This is because if a photon of light hit an electron, it would alter its motion. The photon of light was necessary to find its position. This was the early development of modern **quantum** mechanics.

Louis de Broglie in 1929

Quantum mechanics has become the foundation for understanding the energies in bonding atoms together to form larger molecules. It has been fairly successful in predicting how much energy would be released when bonds between atoms are taken apart. Its impact has been great in several branches of modern science. But as great as it is, there are indications that it still needs to be further developed.

Atomic Models

Positive charges are located within a central nucleus

Rutherford Model, 1909

Electrons are restricted in circular orbits with different energy levels

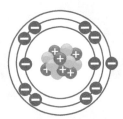

Neils Bohr Model, 1913

Electrons are in clouds surrounding the nucleus, and this cloud is less dense

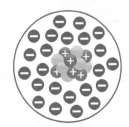

Louis de Broglie and Erwin Schrodinger Model, 1926

Dr. Raymond Damadian was responsible for the concept and creation of what today is known as the MRI, an invention that forever changed the face of medicine.

The better understandings of light have been important in developing many important applications in the use of x-rays, CAT scans, and MRIs.

There is an old saying that the more answers you get, the more fresh questions you get. This is very true. Science is a very important tool that has provided many blessings, but it is still developing. Do not ever feel that everything is understood and there is nothing left for you to do. That is far from true. The creation is so complex that we will never unravel all of its secrets. That is because God is so great — even greater than His creation. Some say that the tools of science are the greatest resource for knowing anything. This is to deny that God even exists. I cannot fathom the greatness of God. That is way beyond my reach. But every time I look at His creation, I realize that there is more beyond it. And to think that He just spoke it all into being. Remember — just because you do not understand something does not mean that it is not true. We do not understand the greatness of God and how He created and maintains everything, but that does not deny it. To the contrary, the sense of discovery and exploration is all the more exciting.

LABORATORY 22

Spectroscopy and Electron Orbitals

REQUIRED MATERIALS

- Spectroscope
- Color paddles
- Penlight
- A flame
- Solar cell
- Galvanometer

Introduction

The photoelectric effect presents the theory that for light to cause electrons to be ejected from certain metals, it has to have at least a minimum frequency. Any frequencies of light below this minimum value will not have enough energy to eject the electrons. This was the major support for the photon model of light. If light were behaving like a wave, it should be gradually absorbed by the metal until it had enough energy to eject electrons. That, however, was not the case. Red light has lower frequency and violet has the greatest frequency in the visible light range. From the low to high frequencies are red, orange, yellow, green, blue, and violet.

Further evidence for the idea that electrons have energy levels, called quantum levels, is the light of specific frequencies given off when electrons go from higher energy levels to lower energy levels. If electrons could have any energy value, the light given off when they went to lower energy levels would be continuous and not broken down into set values. When blue light is given off, electrons are going from higher to much lower energy levels. When red light is given off the electrons are not losing as much energy — not going to much lower energy levels.

Purpose

This laboratory provides experience observing the photoelectric effect that supports the photon model of light. It also provides experience observing the light emitted when electrons go from higher energy quantum levels to lower energy quantum levels, which supports the quantum model of electrons.

Procedure

 observe

1. Connect the wires coming from the solar cell to the galvanometer. The galvanometer will indicate if a current is being produced by the solar cell and the strength of the current. Test your setup by shining a bright light on the solar cell and watch for the deflection of the needle on the galvanometer.

2. Shine the beam from the penlight through each of the colors of the color paddle from red to violet. Write down the results from the galvanometer for each of the colors. The photoelectric effect is that the lower frequency colors of light do not have sufficient energy to eject electrons from the solar cell.

? question

3. Are your results what you would expect? Explain.

research

4. Look through the spectroscope with the end of the tube with the narrow slit pointing toward a bright light (not the sun — too bright). Practice so that you can focus the light onto the black inner wall of the tube. The end you are looking through should have a round window covered with a diffraction grating that will split the light into its various colors. This may take a little practice.

hypothesis

5. When you are able to see the colors focused onto the inner wall of the tube, go into a darkened room and look through the tube at the light from the penlight. While using the penlight as a light source, place the different colors from the color paddle over the end of the tube with the slit so that you are looking at different colored lights. Write down what you observe through the tube for the bright light, light from the penlight, and the different colors of light.

experiment

6. Observe a flame, such as on a gas stove or propane camping burner, through the spectroscope and describe what you observe. If you use a natural gas stove, you are looking at the spectrum produced when methane (CH_4) burns. This should be a more complex pattern because several elements are involved.

analyze

7. Describe what you see. Be careful and keep a safe distance from the flame.

conclusion

8. Do your results support the quantum theory of the atom? Explain using complete sentences.

CHAPTER TWENTY-THREE

RADIOACTIVITY

OBJECTIVES

At the conclusion of this lesson the student should have an understanding of

- The structure of atomic nuclei and isotopes
- The transmutation of elements
- Artificial transmutation of new elements
- Nuclear force
- Nuclear fission
- Applications of nuclear chain reactions
- Detection of radioactivity

A good place to begin the study of radioactivity is to go back to the work of Becquerel in 1896 when he discovered natural radioactivity. This was followed up with the work of the Curies and Rutherford who found that natural radioactivity was actually three different forms of radioactivity — **alpha particles**, **beta particles**, and **gamma rays**. So much has happened since that time. Since the development of nuclear weapons, the question developed — are the scientists responsible for the consequences of their discoveries. At one time it was thought that scientists made discoveries and philosophers decided on the morality of their uses. That has changed. Back in high school, I had the experience of a Spanish teacher telling me that I was evil for deciding to go into the sciences. He said that the only purpose of science was to kill people. That was quite a shock for a high school student, especially since I had taken three years of Spanish with him. The truth is that you can do harm with anything. This is why they are so careful about what they allow inmates to have in prison.

Henri Becquerel in his lab

Radioactivity is a blessing and a curse. It is a two-edged sword. It can cause cancer and it can cure cancer. It is helpful in determining whether the thyroid gland is absorbing sufficient iodine. It can be used to track very small amounts of vital nutrients through the body. It has had many applications in understanding cellular functions that are used to determine health conditions.

To begin with, it is necessary to consider several features of atoms. The idea that most of the space in an atom is empty except for the electrons was studied in the previous chapter. The particles within the nucleus are called **nucleons**. These include the protons and neutrons. They both have essentially the same mass, which is about 2,000 times that of an electron. As the electrons are negatively charged, protons are positively charged, and neutrons have no charge. The total mass of an atom is determined by the number of protons and neutrons, which is called the **atomic mass number**. The identity of an element is determined by its number of protons, called the **atomic number**. For example, any atom with 6 protons is carbon. Any atom with 1 proton is hydrogen. Six is the atomic number of carbon and 1 is the atomic number of hydrogen.

1 proton = hydrogen

Atoms of the same element can, however, have different numbers of neutrons. Atoms with the same number of protons but different numbers of neutrons are called **isotopes**. This is a term of relationship. Someone who is the only child in a family cannot be a brother or sister to someone. Likewise, an isotope has to be compared to atoms with the same number of protons but different numbers of neutrons. Isotopes have the same chemical properties but can have different nuclear properties. Some isotopes of an element can be radioactive while others are not. Back in the 1950s, there was a lot of excitement about the possibilities of radioactivity, most of which were not true. A number of weird movies came out where radiation caused such things as radioactive mud oozing out and melting people, animals growing to wild proportions, and every beast imaginable becoming a monster. These gradually died off as people started to better understand radioactivity.

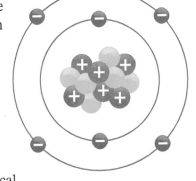

6 proton = carbon

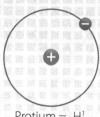

Protium = $_1H^1$

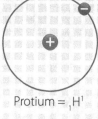

Deuterium = $_1H^2$

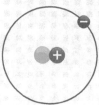

Tritium = $_1H^3$

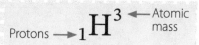

Protons → $_1H^3$ ← Atomic mass

There are 3 isotopes of hydrogen — hydrogen (sometimes called protium), deuterium, and tritium. A hydrogen atom has one proton and no neutron. It has an atomic number of 1 and an atomic mass number of 1. It is symbolized by $_1H^1$. The lower number is the atomic number (number of protons) and the upper number is the atomic mass number (number of protons and neutrons). This is the most common form of hydrogen. Deuterium has 1 proton and 1 neutron making it $_1H^2$ and tritium has 1 proton and 2 neutrons as $_1H^3$. When deuterium combines with oxygen to form water, it is sometimes written as D_2O instead of H_2O. An oxygen atom has an atomic mass number of 16 because it has 8 protons and 8 neutrons. Combined with deuterium, it has a total mass of 2 + 2 + 16 or 20. It is called heavy water. Hydrogen combined with oxygen has a combined mass of 1 + 1 + 16 or 18.

Out of the 3 isotopes of hydrogen, tritium is radioactive. A neutron becomes a proton and an electron (beta particle) as the hydrogen atom becomes a helium atom.

$$_1H^3 \rightarrow {_2He^3} + {_{-1}e^0}$$
$$_0n^1 \rightarrow {_1p^1} + {_{-1}e^0}$$

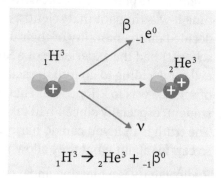

Notice that the atomic numbers add up to the same on both sides of the equation 1 → 2 − 1 and the atomic masses add up the same on both sides of the equation 3 → 3 + 0. The −1 atomic number for the electron means that it has a −1 charge. It can also be written as . . .

$$_1H^3 \rightarrow {_2He^3} + {_{-1}\beta^0}$$

A hydrogen atom became a helium atom — a different element. The new element produced is called the **daughter element**. The process is called **transmutation**. Tritium can be used instead of hydrogen in compounds to be injected into the body and traced by their β emission. In small amounts, it is relatively harmless. β particles have different energies when released by different radioactive elements.

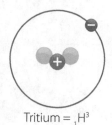

Carbon 11

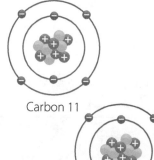

Carbon 12

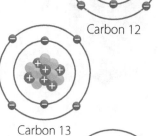

Carbon 13

Carbon has 4 isotopes — $_6C^{11}$, $_6C^{12}$, $_6C^{13}$, and $_6C^{14}$. Out of these, $_6C^{12}$ (called carbon 12) is the most common. The number of protons (atomic number) is always 6 for carbon. The number of neutrons for each isotope is the atomic mass number minus the atomic number. For example, $_6C^{12}$ has 6 protons and 12 − 6 = 6 neutrons. Out of the 4 isotopes of carbon, $_6C^{14}$ is radioactive. Carbon 14 emits a beta particle while becoming a nitrogen atom with 7 protons.

$$_6C^{14} \rightarrow {_7N^{14}} + {_{-1}e^0}$$

Here again a neutron becomes a proton and an electron (beta particle). This electron is not from the orbitals outside of the nucleus but from within the nucleus.

Cosmic rays from the sun and outer space collide with atoms in our upper atmosphere causing them to transmutate, releasing protons and neutrons. The protons (because of their positive charges) combine with electrons to form hydrogen atoms. The neutrons, however, because they are neutral in charge, keep going until they collide with the nuclei of atoms like nitrogen, forming carbon 14.

$$_0n^1 + {_7N^{14}} \rightarrow {_6C^{14}} + {_1H^1}$$

The ratio of carbon 14 in the atmosphere to carbon 12 is about 1 to 1 million. The newly formed carbon 14 combines with oxygen to form radioactive carbon dioxide (CO_2) which can be incorporated into sugar in plants by photosynthesis. This will come up again later in the discussion of carbon dating.

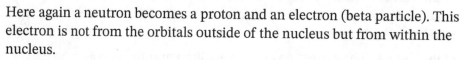

Carbon 14

Another example of natural transmutation is that of uranium 238 ($_{92}U^{238}$). With an atomic mass number of 238 and an atomic number of 92, uranium 238 has (238-92) 146 neutrons. Uranium 238 becomes thorium 234, an alpha particle and gamma rays.

$$_{92}U^{238} \rightarrow {_{90}Th^{234}} + {_2He^4} + \gamma$$

Thorium 234 is also radioactive and becomes protactinium 234 and a beta particle.

$$_{90}Th^{234} \rightarrow {_{91}Pa^{234}} + {_{-1}e^0}$$

Just like in carbon 14, a neutron becomes a proton and an electron (beta particle) in the transmutation of thorium. When a beta particle (electron) is emitted, it is always accompanied by the emission of a neutrino which is a neutral particle with essentially zero mass. As highly energetic particles, they can penetrate almost anything without leaving a trace. They are also emitted by the sun and are zipping right through you right now. They balance the energy on both sides of the equation so that energy is conserved.

Lead 214 decays to form bismuth 214 as . . .

$$_{82}Pb^{214} \rightarrow {_{83}Bi^{214}} + {_{-1}e^0} + {_2He^3}$$

Here a neutron becomes a proton and an electron. The Greek letter nu (ν) stands for a neutrino.

Sometimes when beta particles are emitted, positrons (positive charged electrons) are also emitted. This is the basis for positron emission tomography (PET scans) which can give three-dimensional images of tissues like the brain.

Back in the middle ages, alchemists unsuccessfully tried to transmutate lead into gold. Get rich schemes go way back. As you can imagine, they were unsuccessful. One of the earlier successes in artificial transmutation was when Ernest Rutherford in 1919 bombarded nitrogen atoms with alpha particles.

$$_2He^4 + {_7N^{14}} \rightarrow {_8O^{17}} + {_1H^1}$$

He formed oxygen 17 (the most common form is oxygen 16) and hydrogen. Much later, artificial transmutation was used to form elements that were previously unknown from atomic numbers 93 to 106.

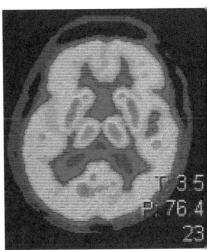

PET scan of brain tissue

You may have been wondering why some atoms are radioactive and others are not. Within the nucleus of an atom, the positive charges of the protons exert tremendous repulsive forces on each other. It would seem like this force would be so strong that atoms would not even exist. Forces are detected by their results. If you hit a baseball with a bat, there is a force exerted on the ball. How do you know? The ball goes flying through the air. It would not have done that by itself; therefore, you exerted a force on the ball with the bat. In the case of the nucleus of an atom, in order for it to hold together, there has to be another force to overcome the repulsion of the protons. This is called the nuclear force. How is it detected? It is there because the nuclei of atoms do not all fly apart. It has to be a very strong force because the repulsions between the protons are strong.

In speaking about Christ, Colossians 1:15–18 states:

> **He is the image of the invisible God, the firstborn over all creation. For by Him all things were created that are in heaven and that are on earth, visible and invisible, whether thrones or dominions or principalities or powers. All things were created through Him and for Him. And He is before all things, and in Him all things consist. And He is the head of the body, the church, who is the beginning, the firstborn from the dead, that in all things He may have the preeminence.**

Nuclear force within the nucleus of an atom

When you hit a baseball with a bat, you exert a force on the ball for a very short time. In the case of the force holding the nucleus of an atom together, it is a force that continues as long as atoms exist. Like the force of gravity, we do not know of any cause of the nuclear force, or the repulsive forces either, other than Christ. Just giving a force a name does not explain its origin or continual source of energy. Many will tell you that these forces can hold the universe together without having to go to any supernatural causes. They will say that to believe that God is the originator and maintainer of these forces is to resort to superstitions. Be careful, because these can sound like strong arguments if you succumb to peer pressure and get your eyes off the Lord.

Because the neutrons have no charge, they are looked upon as increasing the distances between the protons and acting as a "nuclear cement" that helps overcome the repulsions between the protons. Smaller atoms with fewer protons usually are most stable when they have the same number of neutrons as protons. For heavier atoms with many more protons, the repulsions are greater, so they are more stable with many more neutrons than protons. Those atoms without the proper number of neutrons tend to be more unstable and more likely to undergo nuclear decay. They decay by giving off alpha and beta particles. All the atoms in a sample do not decay at the same time. They gradually break down over time. This is where the half life comes into play as will be seen in the next chapter.

Prior to 1932, alpha particles were used to rearrange the nuclei of atoms. An example is this reaction where alpha particles were fired at beryllium atoms. Alpha particles are difficult to work with because their positive charge would be repelled by the nuclei of atoms. They would have to be fired at very high velocities directly at the nuclei of atoms.

$$_2He^4 + {_4}Be^9 \rightarrow {_6}C^{12} + {_0}n^1$$

Neutrons were discovered in 1932, giving physicists a new "bullet" to fire at the nuclei of atoms. That made possible the discovery of **nuclear fission** as shown here.

$$_0n^1 + {_{92}}U^{235} \rightarrow {_{56}}Ba^{142} + {_{36}}Kr^{91} + 3\ {_0}n^1$$

Fission means to break apart. Notice that the neutron "bullet" hitting the uranium nucleus broke the uranium nucleus apart and resulted in the release of 3 new neutron "bullets." This introduced new possibilities. Uranium found in nature is 99.3% uranium 238 and about 0.7% uranium 235. There is a slight mass difference between the two forms of uranium. Their different densities is the basis for their separation. If the uranium atoms could be concentrated enough, a **chain reaction** could be produced. This would cause many more uranium atoms to undergo fission, releasing tremendous amounts of energy. The chain reaction feeds itself because 3 neutrons are produced for every one used in the reaction. This is the basis for its use in a bomb or a nuclear reactor to produce electric energy. In nuclear reactors there have to be graphite (carbon) moderators (water is also used as a moderators) and boron control rods to slow down most of the neutrons to keep the chain reaction from getting out of hand. The heat from the fission reaction heats water to steam which turns giant turbines that generate high voltage electric currents. The turbines are like large paddle wheels that rotate magnets that induce electric currents. The reaction of neutrons with uranium 235 produced 3 neutrons, 2 of which had to be absorbed by the boron control rods so that the reaction would be self-sustaining at the same level — one neutron in and one neutron out.

One of the byproducts of the process is the production plutonium 238 formed when neutrons interact with the uranium 238 that is mixed in with the uranium 235. Plutonium 238 was stored up to make nuclear bombs near the end of World War II. When the bombs were dropped upon Japan, history changed. This was why my Spanish teacher considered science to be evil. Ever since then, science has not been without moral obligations for its discoveries.

How are alpha particles, beta particles, and gamma rays detected and measured? Remember that when radiation is given off it goes out in all directions like light from a bulb. One way to control the pathway is to generate it in a lead container with a very small opening.

One of the early means of detection was a Geiger counter. This is metal cylinder with a central metal rod. The opening to the cylinder is a window of a very thin layer of metallic mica that allows radiation to enter the cylinder. The inner metal layer of the cylinder has a negative charge and the central rod has a positive charge from a power supply. A gas in the cylinder is almost ionized (able to lose outer electrons) by the voltage between the cylinder and the rod. When radiation enters the front window it ionizes some of the gas molecules, sending electrons to the positive rod and positive ions to the negative inner wall of the cylinder. This temporarily completes the circuit which triggers a counter. The results are in counts per minute (CPM). The more radiation, the more counts. This does not count all of the radiation — only those that enter the front window. This is a relative rather than absolute number. It tells it there is a lot or little radiation (*Diagram 23.1*).

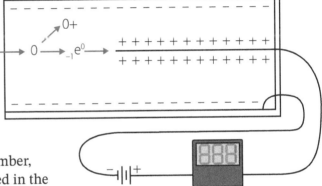

Diagram 23.1

A second tool for detecting radiation is a cloud chamber. It is a glass cylinder with water or alcohol vapor with a glass window at one end and a moveable piston at the other end. When the piston is dropped, suddenly increasing the volume in the chamber, the vapor is cooled. When a radioactive sample is placed in the chamber, radiation produces ions (charged atoms or molecules) in the cooled vapor similar to what happens when a jet passes through cool air at higher altitudes, producing ice crystal trails. The trails can be photographed. When a strong magnet is placed around the chamber, positive (alpha) particles and negative (beta) particles pathways are bent in opposite directions so they can be distinguished from each other (*Diagram 23.2*).

A third tool is a **bubble chamber**, which is a glass chamber with liquid hydrogen just under the boiling temperature of the liquid hydrogen. If the pressure is quickly lowered by increasing the volume just as the radiation enters the chamber, a trail of bubbles is left in the radiation's trail. Photographs are taken quickly just before the hydrogen boils. The pathways left behind reveal the charge and relative mass of the particles.

A fourth tool is a scintillation counter. Some compounds give off light when charged particles pass through them. When the radiation passes through one of these compounds, the light flashes are counted with light detectors and converted to electric signals which go to counting devices. Many times these are more sensitive than Geiger counters.

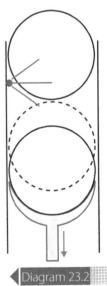

Diagram 23.2

The applications of nuclear energy have produced many blessings from the generation of electrical energy to treatments of cancer. But as with electricity, nuclear energy is also a blessing and a curse. It demonstrates the benefits of taking the time and effort to understand it and carefully develop its many applications. In applying all of God's blessings, we have to be diligent to carefully study and apply what He has provided.

LABORATORY 23

Radioactivity

REQUIRED MATERIALS
- Watch or clock with a luminous dial (if you have one or can borrow one)

Introduction
When alpha particles strike the sulfur atoms in zinc sulfide, a flash of light is emitted. This is the basis behind a scintillation counter. The counter is an instrument for detecting and measuring ionizing radiation by using the excitation effect of incident radiation on a scintillating material and detecting the resultant light pulses given off per minute.

In nuclear reactions, the number of nucleons are conserved. Step 2 provides practice in balancing these reactions.

There are many beneficial uses for radioactivity as well as harmful uses. In a fallen world, there are nuclear hazards in the environment. It would be interesting if we could see the roles of radioactivity before the Fall. A topic of current research is from the view that everything has beneficial purposes. The assigned paper is an exercise in investigating the beneficial purpose of radioactivity of your choosing.

Purpose
This exercise is to provide practice in seeing an example of the detection of radioactivity, balancing nuclear reactions, and researching a beneficial use of radioactivity.

Procedure

 observe

1. If you have or can borrow a luminous watch or clock take it into a closet or darkened room (or at night) and look at the dial after your eyes have adjusted. Using a magnifying lens (perhaps a lens from lab 18), look very closely at the glowing dials. You should be able to see tiny flashes of light that normally blend together when you are

not using the lens. The dials have very small amounts of radium bromide that give off alpha particles that glow when they strike zinc sulfide.

? question

2. What do you see? If you cannot get hold of a watch with luminous dials, you can skip this part.

research

3. It may not be a good idea to be having radioactive materials around the house or having to go out and buy a Geiger counter, so for this exercise you are going to practice working with some of the symbols and reactions from this chapter. As you do these, you can go back and look in the chapter. Be a good detective.

hypothesis/experiment

4. How many protons are in this atom — $_{42}Mo^{96}$?

5. How many neutrons are in this atom — $_{79}Au^{197}$?

6. How many neutrons are in this atom — $_{90}U^{238}$?

7. $_{11}Na^{23} + _{2}He^{4} \rightarrow _{1}H^{1} + \_\_Mg^{\_\_}$

8. $\_\_Be^{\_\_} + _{2}He^{4} \rightarrow _{0}n^{1} + _{6}C^{12}$

9. $_{7}N^{14} + _{2}He^{4} \rightarrow _{1}H^{1} + \_\_O^{\_\_}$

10. $_{90}Th^{234} \rightarrow \_\_Pa^{\_\_} + _{-1}e^{0}$

analyze/conclusion

11. Write a one-page double-spaced 12 font paper on one of the beneficial uses of radioactivity. If you can, choose an application that you are more familiar with or has been used to benefit someone you know. It will be graded on grammar, proper sentence structure, spelling, and how clearly it communicates. Cite the sources that you use on a separate page, following your teacher's instruction as to how to cite the source. It should have enough information to give proper credit to the source and enable anyone to go back and find it.

CHAPTER TWENTY-FOUR
APPLICATIONS OF RADIOACTIVITY

OBJECTIVES

At the conclusion of this lesson the student should have an understanding of

- The exposure of radioactivity to living tissue
- Radiocarbon dating
- Radiometric dating with radioactive metals
- Nuclear fusion

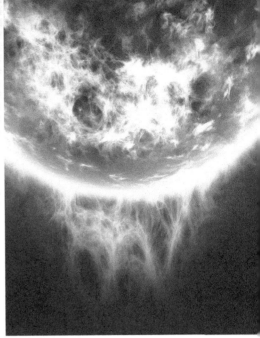

When radiation enters living tissue, several different things can happen, depending upon what organelles it encounters. If DNA is hit, it can cause breaks which can produce mutations which can be harmful, or if it hits a gene that is involved in the regulation of cell division, cause cancer. When radiation hits other tissues and molecules within cells or extracellular fluids, it can produce **free radicals**. These are atoms within molecules that have unpaired electrons that otherwise would have been paired with another electron. Free radicals are very reactive, causing damaging reactions with essential molecules. Antioxidants in the diet will help reduce free radicals.

Even though alpha particles are larger and do not travel as far as beta particles, gamma rays, and X-rays, they can be more damaging. Alpha particles are more likely to produce several damaging events in a short space of each other. This can be more harmful because they can do several things to one cell, whereas beta particles, gamma rays, and x-rays are more likely to spread their effects out over several cells.

Cells have a remarkable ability to produce enzymes that repair damage from radiation. This is grace that was provided in the DNA of the original life forms of creation. If a cell is extensively damaged, it can be digested by white blood cells and enzymes and replaced. This is true for most cells, except for those in the central nervous system, which cannot repair themselves.

An average person in the United States gets about 56% of the radiation from background radiation from minerals in the earth and cosmic rays from the sun and outer space. About 42% of the radiation that we are exposed to over our lifetime come from medical x-rays. This percentage is higher for those who have radiation treatment for cancer. About 2% comes from radiation in the atmosphere from previous weapons testing fallout.

Radiation dosages are measured in **rads** (radiation absorbed dose). This is the amount of energy absorbed per gram of exposed material. Human exposure is measured in **rems**, which refers to Roentgen equivalent man. It is found by multiplying the number of rads by a number that reflects the overall effects of a specific type of radiation. For example, one rad of slower alpha particles has the same effect as 10 rads of faster beta particles, gamma rays, and x-rays. The one rad for alpha particles is 10 rems and 10 rads of beta particles; gamma rays and x-rays are also 10 rems. The average person in the United States is exposed to about 0.2 rems a year. Those closer to the poles are exposed to a bit more because there is not as much shielding from earth's magnetic field there. About 500 rems are lethal. We are probably more at risk from skin cancer due to ultraviolet radiation from the sun than from radioactivity.

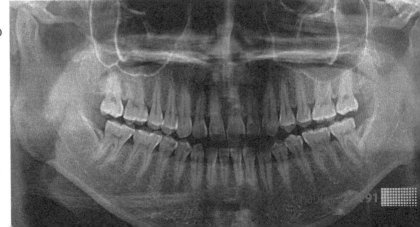

An important application is that of radiometric dating. It is a major factor in many Christians deciding that the earth has to be very old. Interestingly, how the flood in the days of Noah is viewed has a tremendous bearing on this topic. If the earth is very old, most of the fossils buried in the earth would have been buried over long periods of time. If the flood was worldwide, most of the fossils would have been buried in a very short period of time leaving less geological evidence for an old earth. This brings the debate back to Genesis. I have been told by a leading Hebraist (Dr. Steven Boyd) that the style of Hebrew used in the early chapters of Genesis are historical and not meant to be taken figuratively. He said that this would be obvious to a 12-year-old Jewish boy. The worldwide flood described in Genesis would account for the fossilization that we find because a flood of that magnitude would reshape the surface of the earth. To go with an old earth, one has to look upon the flood as being localized to that part of the world and treat Genesis as figurative. Here the physics of radiometric dating become very important. Does radiometric dating leave no room for a young earth? In looking at this topic, we need to be objective as we have done with the rest of the concepts of physics in this book.

Radiocarbon dating deals with C^{14} atoms that are formed in the atmosphere and incorporated into plant tissue as CO_2 by photosynthesis. Animals eat plants, which incorporate it into their tissues. At the time a plant or animal dies and is buried, the ratio of C^{14} to C^{12} in their tissues is the same as that in the atmosphere. Carbon dating is, therefore, used for fossilized life forms or other objects with carbon. This is accomplished using the concept of the half-life of C^{14}. The **half-life** is the time it takes for half the sample of C^{14} to break down into N^{14}.

$$_6C^{14} \rightarrow {}_7N^{14} + {}_{-1}e^0$$

The half-life of C^{14} has been calculated to be 5,730 years. If I told you that I had 1 candy bar and I would give each of 30 students half of the candy bar, you would think twice. I would give one of you half, the next half of the half left, and the next half of what was left over, and so on. That means that the first one gets half, the next actually gets one-fourth, and the next one-eighth, and so on. This is what a mathematician calls an exponential function. This means that when an animal dies and it stops taking in carbon, its C^{14} is no longer being replaced. After 5,730 years, the animal's bones would have half as much C^{14} left. After 2 half-lives, 11,460 years, it would have $\frac{1}{4}$ as much of the original C^{14}. After 3 half-lives, 17,190 years, it would have $\frac{1}{8}$ as much of the original C^{14}. If the amount of C^{14} compared to the amount of C^{12} is measured in a bone fossil, it could be calculated as to how long the bone had been buried. If the flood happened only 4,350 years ago as has been estimated, a bone over 17,000 years old could not have been buried by the flood.

One of the first tests of a scientific hypothesis or conclusion is to examine the assumptions that went into it. C^{14} forms in the upper atmosphere by cosmic rays causing transmutations of many molecules in the upper atmosphere. Some of these result in releasing neutrons that convert nitrogen atoms into C^{14} atoms and hydrogen atoms.

$$_0n^1 + {}_7N^{14} \rightarrow {}_6C^{14} + {}_1H^1$$

The amount of C^{14} assimilated into plant and animal tissue in the past depends upon the amount of C^{14} in the atmosphere at that time. It has been shown that the amount of cosmic rays have varied in the past because the magnetic field of the earth has varied in the past. As well, a lot of carbon is stored in the oceans. Colder water releases less CO_2 into the atmosphere than warmer water. Variations in the temperature of the oceans in the past will vary the amount of CO_2 containing C^{14} released. This means that the C^{14} incorporated into plant and animal tissue in the past has not been constant. Some studies have tried to take this into account and recalibrate the amount of C^{14} in the atmosphere in the past. This is commendable but relies upon knowing with certainty how the conditions in the past varied.

This is difficult to know with certainty because the means to measure ocean temperatures and cosmic rays were not available very far back in time.

One of the best tests is to test samples that were buried with written materials or other means to establish when a fossil was buried and compare that to the radiocarbon date. Studies have shown C^{14} dates to match records verifying the dates go back as far as 400 B.C.

Verifying the dates of samples is different than the accuracy of measuring the C^{14} content. Most say that if a sample is 50,000 years old or younger, there is enough C^{14} to measure. Laser measuring technics are being designed that could measure even smaller amounts of C^{14} extending it back to about 100,000 years. Remember that being able to measure C^{14} and to establish a date are two different things.

Diamond

There are fossils, coal deposits and diamonds that are considered to be over a million years old that should not have any C^{14} left, that do still have measurable C^{14}.

Another important assumption is that the half-lives have been constant over large spans of time. Without this, the technique would only be as good as the half-lives remained constant.

The take home lesson here is that radiocarbon dating methods can only be supported back to about 400 B.C. It does not have to be considered that a young earth view is deemed invalid from reliable research data. Our conclusions are only as reliable as our data, and the assumptions that go into interpreting the data. Many Christians do not accept a young earth interpretation of Genesis because of radiometric dating. This does not have to be the case when we objectively look at what we have to work with. Genesis and references are very clear about a flood that covered the earth, and with an old earth view that is not possible because it would have to have been a flood that did not leave many layers of fossils. A flood of that magnitude is going to totally reshape the landscape and bury everything that was there.

In spite of the dating questions, radiocarbon dating has been a tremendous tool for short-term dating. It has helped in understanding many of the remnants of past civilizations and filling in a lot of gaps for archeology.

Another important application is that of uranium dating and using other radioactive heavy metals. These are used where rock samples do not bear carbon. They are also applied using samples in the same rock as fossils. Many of them have very long half-lives. For example, U^{238} becoming Pb^{206} (lead 206, this occurs over several steps) has a half-life of 4.47 billion years. This is close to the uniformitarian estimate of the age of the earth at 4.6 billion years. Using this half-life, some rocks (Precambrian — lowest layer of sedimentary rock with little or no fossils) date back to over 4 billion years old. As with any

Precambrian rock layer

technique in science, the assumptions that went into it have to be considered. This is true for any technique; they all have assumptions. One is that the initial conditions of the sample are known. For example, were some of the daughter atoms present when the rock first formed by being buried as sediment? The rock could have been formed as cooled volcanic magma, but in this case, there would not have been any fossils. A technique called the isochron method has helped answer some of these questions.

Another assumption is that there has not been any exchange of materials with the sample over the time span that it was buried. If some of the parent or daughter isotopes were leached out or washed into the sample, it would change the date of the rock. Some of this is also addressed by the isochron technique.

A third assumption is that the half-life has been constant for the entire time that the rock sample was buried. This is usually treated as a constant, like the gravitational constant and the speed of light. The half-lives are calculated from the decay rate over a short period of time and then calculated back to when half of the sample was present. The math used is an exponential equation that requires a bit more math background. If this were true, the earth could not have an age in the thousands. This is one reason why many have concluded that the Genesis creation account cannot be taken literally for its details and that the flood account was not worldwide. If the earth were very old, then the fossils would have been formed over vast spans of time and could not have been formed rapidly in a worldwide flood. But if the half-lives have changed over time, the earth could be much younger. Is there any evidence of a changed half-life? A study reported in the journal *Science* in 1999 reported where several German scientists stripped all 187 electrons from a rhenium 187 atom and its half-life dropped from 42 billion years to 33 years. It appears that the electrons were shielding the nucleus and holding back the nuclear decay process. This was a very high energy unusual event that could only occur naturally in the intense heat of a **star** rather than on earth. It does demonstrate that the half-life could vary. In any scientific study, it is important to always consider new options and hypotheses. This is how our understanding and knowledge grows. The basic lesson here is that nuclear half-lives do not require one to abandon the literal interpretation of Genesis.

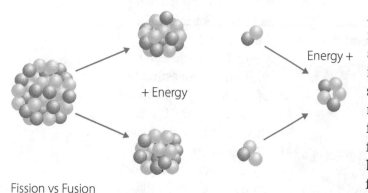

Fission vs Fusion

Another application of radioactivity is that of nuclear fusion. Nuclear fission is splitting nuclei apart and fusion is putting them together. This is an ongoing process in stars, including our sun, where hydrogen nuclei fuse to form helium nuclei. Iron atoms and those smaller, undergo fusion and those heavier than iron undergo fission. If it takes so much energy to fuse hydrogen atoms, it would take so much more to fuse heavier nuclei with more protons.

When protons are brought close together, their repulsion (because of their positive charges) increases tremendously. They have to be forced together by an enormous amount of energy, and when they do come together, the energy of repulsion is released as thermonuclear energy. This is the basis of the hydrogen bomb.

$$_1H^2 + {}_1H^3 \rightarrow {}_2He^4 + {}_0n^1$$

The total mass on the right side of the equation is a little less than the mass on the left side. It has been converted into energy according the relationship discovered by Albert Einstein where $E = mc^2$, where m is the missing mass and c is the speed of light.

This reaction has to occur in a small space at very high temperatures — more than a million degrees Celsius. A fission bomb is exploded to provide the heat and to push the H atoms together. This is why it is called a thermonuclear bomb. The fusion explosion then releases more energy than the fission explosion.

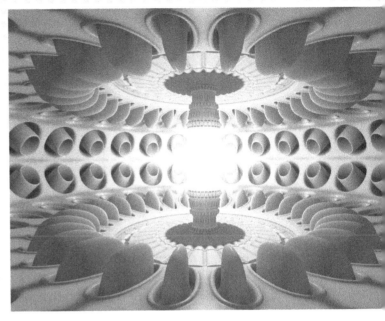

Fusion reactor core

Current research involves how to control the energy released from **nuclear fusion** to reduce sewage and garbage and industrial waste to its basic atoms (fresh raw materials), desalinate sea water, produce electricity, etc. That energy could be very destructive if not controlled. In nuclear reactors, the temperature released would melt any solids, so magnetic fields have laser beams that heat a very small confined pellet of fusion fuel. In 1988–1989, scientists in Utah claimed to have developed cold fusion reactions that did not require high temperatures or high densities. They lowered a palladium electrode into a beaker of heavy water (D_2O). When they passed an electric current through the electrode, that drew deuterium atoms into spaces between the atoms of the palladium. They produced heat that could not be explained by chemical and physical causes. Some saw this as a tremendous breakthrough for a cheap and abundant energy supply. Several tried to repeat the experiment but did not succeed. It has not been repeated since. In order to validate their claims, someone else would have to be able to repeat the experiment and get the same or very similar results. In science, it is very important that an experiment be repeatable for validation and further study. Evolution, by its very nature, is not repeatable. Neither is creation. God has preserved knowledge of how He created in His sovereign Word. That is only right, because His word is of the highest authority.

Laboratory 24

Applications of Radioactivity

REQUIRED MATERIALS

- 2 Bar magnets
- Graph paper

Introduction

Nuclear fusion involves combining the nuclei of small atoms overcoming the strong repulsion of their positively charged protons. It requires a very small space and enormous temperature. It is simulated in this exercise on a much smaller scale with 2 bar magnets.

A nuclear chain reaction multiplies itself very rapidly, releasing enormous amounts of energy. It increases in magnitude as it progresses. It is simulated here in a harmless manner with numbers in squares of graph paper.

Nuclear weapon test that was part of Operation Castle, which began in March 1954

Purpose

This exercise simulates, on a much smaller scale, the repulsion to be overcome in a nuclear fusion reaction and the magnitude and rapid expansion of a chain reaction.

Procedure

 observe

1. Take 2 magnets that you used in lab 16 and try to push them together so that the same poles face each other. Push them together carefully. Describe what it feels like as they get closer to each other.

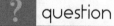 question

2. Are you able to get them to actually touch each other without moving to the sides? This is similar, but nowhere near as strong as that of the repulsion between the protons in two nuclei being pushed together in nuclear fusion.

research

3. Take a sheet of graph paper with at least 10 squares on each side giving a total of at least 100 squares.

   Choose a square in the middle and place the number 1 in it. Choose 3 squares around and near the first one and place a number 2 in each of them.

   Choose 3 squares around and near each of the squares with a number 2 in it and place a 3 in each of them. How many squares have a number in them?

   Choose 3 squares around and near each of the squares with a number 3 in it and place a 4 in each of them. How many squares have a number in them?

   Choose 3 squares around and near each of the squares with a number 4 in it and place a 5 in each of them. How many squares have a number in them?

 hypothesis

4. List the total numbers of squares with numbers from the above steps.

 experiment

5. Describe the sequence of numbers. This is how a chain reaction works.

 analyze

6. What you could have done — but it would have asked for a bit much — is to set mouse traps all over the floor of a gymnasium with a ping pong ball on each mouse trap. Then stand back and throw ping pong balls, one after another, onto the mouse traps. I think the graph paper is more reasonable.

 conclusion

7. How would you describe this experiment overall?

CHAPTER TWENTY-FIVE

TIME

OBJECTIVES

At the conclusion of this lesson the student should have an understanding of

- The principle of relativity
- Special relativity
- General relativity

We have used the concept of time (as measured by clocks) in discussions of velocity and acceleration which we use every day. We get up at certain times. We go to bed at certain times. Some can even tell how much time has gone by without looking at a clock. But our experience does not involve velocities close to the velocity of light, nor do we live under immense gravitational fields as found near large stars and galaxies. That is more the topic of this chapter. It is a bit mind-expanding, but just as real as our everyday thoughts of time. Like other aspects of creation, God, not being part of creation, is not held by the laws restricting creation. In Scripture, especially in the Old Testament, there are occasions where Christ appears way before being born in Bethlehem. Theologians call these appearances theophanies. When the Jewish leaders asked Christ who He was, He said "I AM," which is how God identified Himself to Moses at the burning bush.

You have used the concept of time throughout your life up to this point. You had to watch your time. Get up at a certain time. Plan how much time you need for certain tasks that may take days or months to finish. Physicist and atheist Steven Hawking said that man will never build a time machine because we have never been visited by tourists from the future. This chapter deals with different rates of time, but in all of the situations time keeps going in one direction and we cannot stop it. That is part of God's sovereignty. Scripture reminds us that this is the time of salvation, but that the end will come for opportunities to accept Christ's provision of salvation. As you go through life there come major changes and you have to go forward and cannot go backward. You hear people talk about the good old days. It is better to trust God and follow Him today, because He has designed time to pass and go in only one direction.

A note as we discuss relativity. In physics, the concept of relativity has no connection with what some have called situation ethics where right and wrong are relative to the circumstances.

Principle of Relativity

When you think of time, you probably think of how long something takes or when can you do something. To begin the discussion, we need to first consider what is meant by an **inertial frame of reference**. When you measure length, you have to have a ruler of some kind. When you measure time, you need to have something as well, like a clock. The word *inertia* goes back to Newton's first law of motion — that objects at rest tend to stay at rest and objects in motion tend to stay in motion unless there are opposing forces. This is because an object's inertia depends upon its mass. A frame of reference is the background against which you measure something. It could be the ground or floor and walls of a building. If I say that I walk two blocks, I need to say where I began and where I ended up. Maybe I start at the front door of my home and end up at a street corner two blocks away. That is my frame of reference.

If I say that I am going up 2 feet, does that mean that I am going straight up in the air or that I am going up on the steps of a staircase? Remember that speed has quantity but no direction. Velocity has quantity and direction. But where are you starting from and where are you ending up? This is the frame of reference.

Galileo and Newton both agreed that the laws of physics are the same in any inertial frame of reference. What does that mean? This simply means that Newton's laws of motion apply in any frame of reference. They would apply on a beach in Australia and on a submarine under the ocean.

If we are riding in a car, or maybe a horse-drawn carriage as in their day, and I hand you a pencil, the velocity of the pencil is measured in the car, not from the viewpoint of someone standing on a street corner watching us. If the car is going 50 miles an hour down the road, we do not think of the pencil as going 50 miles an hour. The inertial frame of reference is the car — not on the street corner. I can apply the laws of motion in the car without regard to the outside world. This is called the principle of **relativity**. You measure something relative to something else. This is basically the way we have measured things all our lives.

Special Relativity

In his paper on **special relativity** in 1905, Albert Einstein began with the principle of relativity as his first postulate. A **postulate** is an assumption that is made at the beginning of a reasoning process. He called it special relativity because he was referring to how things are measured when the frame of reference was going at tremendous speeds (such as close to the speed of light). That would be a very special circumstance, so he called it special relativity.

His second postulate was that the velocity of light in a vacuum is the same in any inertial frame of reference, regardless of the relative motion of the source and observer. If you measure your velocity while walking down the aisle of an airplane, you would add your velocity in the airplane to the velocity of the airplane to get your total velocity in relation to the ground. But if you were to shine a light on the airplane, you would not add the velocity of light to that of the airplane. The airplane is going so much slower than light, that it would not matter. So the velocity of the light beam would be the same whether you measured it on earth or on board an airplane. Either frame of reference would give the same answer, which is what Einstein meant by the velocity of light being the same in any inertial frame of reference. That makes light rather unique.

Einstein devised what he called **thought experiments** to test the consequences of the second postulate. First, he came up with a way to measure a unit of time with light (instead of your smart phone). He imagined that, aboard a spaceship traveling through space, he shined a light beam that hit a mirror and bounced back. The time for the light beam to go back and forth was his unit of time. He called the distance between the light source and the mirror d, and t_0 to be the time it took the light beam to go to the mirror and back. This would make the velocity of the light beam to be $c = \dfrac{2d}{t_0}$ (velocity is distance divided by time). He used t_0 for time instead of t because that was his frame of reference.

Then the time t_0 would be $t_0 = \dfrac{2d}{c}$.

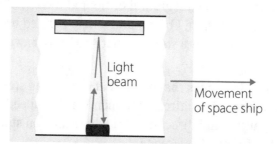

Someone looking from earth would see the pathway of the light beam as a triangle instead of a straight line.

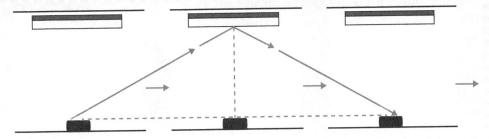

If v is the velocity of the spaceship and t is the time for the light beam to go to the mirror and back as seen from earth, t would be . . .

$$t = \frac{t_0}{\sqrt{1 - \left(\frac{v}{c}\right)^2}}$$

t would be a bit longer than t_0 because it is traveling up and down the triangle instead of just up and down. The equation above was found using geometry of the triangle. This means that the same event takes a little longer as seen from earth as compared to as seen from aboard the spaceship. This is called **time dilation** or expanded time. The equation above is the time dilation formula; t_0 is called the proper time because it is the time as measured in the same inertial frame of reference as the event took place. In other words, the time for the same event is different depending upon the frame of reference where it is observed. The v (velocity of the spaceship) would have to be almost the speed of light for t to be noticeably greater than t_0.

The distance the spaceship traveled in the time it took the light beam to go back and forth would be $L = vt_0$, where v is the velocity of the spaceship and t_0 is the time of the light beam as measured on board the spaceship. L (length) is the distance the light beam traveled. An observer on earth would measure the distance the spaceship traveled as being $L_0 = vt$. L stands for length or distance. L_0 refers to the resting length because the earth is the frame of reference for the observer on earth. Because t is greater than t_0, L is greater than L_0. The distance the spaceship traveled, while the light beam made one up and down movement, is shorter as viewed from earth. This is called **length contraction**. It can be expressed by the equation shown below which is the length contraction formula. Again, the spaceship would have to be going almost the speed of light.

$$L = L_0\sqrt{1 - \frac{v^2}{c^2}}$$

Where do you find anything that travels close to the speed of light so that these equations can be tested in the real world? Earlier we discussed subatomic particles (neutrons, beta particles, alpha particles) that do approach the speed of light. There are smaller particles that make up protons and neutrons that have also been tested. These have given results that show time dilation and length contraction as predicted by these equations.

One important assumption in Einstein's Theory of Special Relativity is that neither the observer on the ground nor the observer in the spaceship are accelerating — their velocities are constant. If they were changing, we could not use the values v and v_0 as unchanging.

An interesting story that is always told in regard to special relativity is that of twins, where one travels to a distant star at close to the speed of light and the other remains on earth. If the one travels at a velocity of 0.995 c, the time dilation gives a factor of 10 for the twin that remains on earth. If it took the traveling twin 10 years to go and come (as measured by the traveling twin), the time for the twin on earth would have dilated (spread out) over 10 × 10 or 100 years. All of the clocks and biological processes for the traveling twin would have been as though only 10 years went by. But when the twin returned, the twin on earth was 100 years older. Of course, this has never been tested, but experiments with smaller subatomic particles have given results that are consistent with the basic idea.

General Relativity

Ten years after he published his work on *special* relativity, in 1915, Einstein published his work on **general relativity**. Special relativity is called special because going at velocities close to the speed of light are special cases. However, gravity, including immense gravities of whole galaxies, are around all the time. So these effects of gravity are called general relativity.

Diagram 25.1
Rotating space station outward centrifugal force is artificial gravity

Special relativity assumes zero acceleration, but general relativity occurs with acceleration. You always have the acceleration of gravity with gravity. If you took a scale to weigh yourself into an elevator, you would see that you weighed more when the elevator accelerated upward, and you would weigh less when the elevator accelerated downward. This is Einstein's principle of equivalence. He said It is impossible to distinguish the acceleration of a frame of reference from the effects of a gravitational field. In other words, acceleration is acceleration, whether it is caused by gravity or something else. In a rotating wheel space station, centripetal force acts the same as artificial gravity toward the inside of the outer wall of the space station (*Diagram 25.1*).

If you toss a ball while standing on the ground, the ball will take a curved path toward the ground with a downward acceleration of $9.8 \frac{m}{s^2}$ because of gravity. If you are in a spaceship with an acceleration of $9.8 \frac{m}{s^2}$ in the same direction, the ball will go toward the bottom of your chamber in the same way. This is why the acceleration of gravity is equivalent to acceleration produced by other means.

If you are in a very rapidly accelerating spaceship, a beam of light will be bent downward. Because gravity and acceleration give the same effect, Einstein reasoned that gravity would bend light. When radio signals (also electromagnetic radiation) from the *Mariner* spacecraft orbiting the planet Mercury were being sent back to earth as Mercury went around the back side of the sun, taking *Mariner* with it, signals kept coming to earth because the strong gravitational force of the sun bent the radio signals going around part of the sun. The time dilation and length contraction of special relativity appeared to be operating in large gravitational fields as well (*Diagram 25.2*).

Gravity produces acceleration, so an accelerated clock will run slower (time dilated or stretched out) compared to a non-accelerated clock. This also produces a gravitational **red shift** where light has a lowered frequency (diluted energy, if you will), making its frequency to be shifted toward the lower energy (red) side of the spectrum. This is where all of the earlier discussions of velocity, acceleration, and light spectra can pay off. You may want to take a minute to review some of these in earlier chapters. When you study, study to acquire tools to help you later rather than just getting by for the time being.

Space and time are tied in with each other and cannot be separated. We measure space by distance and time by how distance is traveled (velocity and acceleration). So, whatever you do to one affects the other. In working with these ideas, Einstein found that Euclidean geometry (that which was developed by Euclid that you study in high school) did not work well with general relativity discussions. He used what is called non-Euclidean geometry. In Euclidean geometry, two parallel lines never meet. In non-Euclidean geometry, they do. This is like the lines of longitude on a globe that meet at the poles.

He realized that if he was going to consider time along with space, he had to use four dimensions — length, width, depth, and time. A Euclidean graph with parallel lines would not work. He used a non-Euclidean graph such as is used to show the gravitational effects of a black hole. This is similar to the globe. Black holes represent huge gravitational fields thought to be caused by the collapse of a massive star. If you flew into a star, you would burn up. But if the mass of the star collapsed into a singularity in its center of mass, you could be drawn in closer before becoming a crispy critter. A black hole has two parts — a singularity in the middle and an outer event horizon, which is the point of no return. When anything comes closer than the event horizon, gravity would be so strong that not even light could escape. A graph showing the effects of the strong gravitational pull of a black hole is shown bent in toward the singularity (*Diagram 25.3*).

Strong gravitational effects of a black hole

Diagram 25.3

Toward singularity

The curvature of this graph shows the effect of time on the other three dimensions. These seem like very strange ideas because we do not live in such strong gravitational fields. It further demonstrates the immense strength and wonder of God's creation.

General relativity states that clocks at low altitudes (stronger gravity) should tick slower than clocks at higher altitudes (less gravity). This is referred to as gravitational **time dilation**. An atomic clock at the Royal Observatory in Greenwich, England, is five microseconds slower each year than the atomic clock at the National Bureau of Standards in Boulder, Colorado. Both clocks are accurate to a microsecond each year. The difference is predicted by the time dilation formula from their difference in gravity, caused by a mile difference in altitude. The significant point here is that a hypothesis is made and then tested experimentally. When the prediction matches the outcome, the hypothesis is supported. This is important because it gives credibility to the value of the hypothesis but also keeps the possibility of improvement open. Einstein himself entertained options that would change his theories. This is different than revelation from the all-knowing God. We do not have to rewrite His Word because it does not need to improve.

Atomic clock in Greenwich, England

LABORATORY 25

Time

REQUIRED MATERIALS

- Biconvex lens (50 mm or 150 mm)
- Biconcave lens
- Laser pointer
- Flat mirrors (as many as you can find)

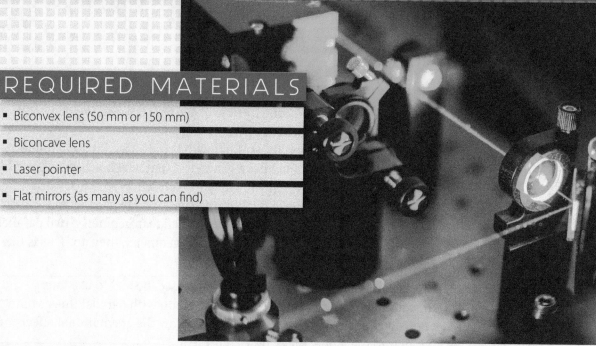

Introduction

The high velocity of light is hard to visualize from the numerical velocity. There are ways to gain a better conceptual feel for the contrast of the greater velocity of light.

Large gravitational fields can act as lenses, changing the direction of light from distant sources. In acting as a lens, gravity bends light rays. The light beam does not return to its original direction as it does when it passes through a different medium.

Purpose

This exercise is to give a feel for the magnitude of the velocity of light and the effect of lens on light beams to simulate the effect of gravitational lenses of large gravitational fields of huge galaxies.

Procedure

 observe

1. This exercise is to help you get a feel for the speed of light. First, see how many flat mirrors (small ones are fine) you can find. You do not need to go out and buy them. You can get by with one, but more would be better.

2. In a large room or back yard, set up a mirror so that you can shine the laser pointer at it and have the beam come back to you. Do not point it at anyone, including yourself, or automobiles or such. If you have more than one mirror, set up the second one so that the light beam goes back to the first mirror. See if you can set it up where the beam goes back and forth. This can be tricky and requires patience.

question

3. How would you describe your observations.

research

4. Now set up the second mirror so that the beam reflects back off center from the first mirror where you can set a third mirror if you have one. Set up more mirrors, if you have them so that you have a pattern of the beam going back and forth.

5. Make a drawing of how you set up the mirrors and describe the results. Again, the idea is to get a feel for how fast light is. The laser pointer is used because it is easier to see and keep track of.

6. Shine the laser beam through a convex lens from lab 18 and project the pattern onto a wall or similar structure. Describe what you see.

7. Shine the laser beam through a concave lens from lab 18 and project the pattern onto a wall or similar structure. Describe what you see. There is a possibility that large nebula and huge gravitational fields in space could act as lenses.

hypothesis

8. Point the laser pointer to a 90° angle to the side of a large glass of water. Observe and describe the laser beam as it enters the glass of water and exits the other side of the glass.

experiment

9. Point the laser beam at a 45° angle to the side of the glass of water and describe what you observe.

analyze

10. Light slows down (which bends the light beam) when it passes through a different medium, and returns to its original velocity when it exits the medium.

conclusion

11. When gravity bends light, it is not going through a different medium, so it does not return to its original direction.

CHAPTER TWENTY-SIX

THE SOLAR SYSTEM

OBJECTIVES

At the conclusion of this lesson the student should have an understanding of

- The solar system
- Astronomical units
- Mercury
- Venus
- Earth
- Mars
- Jupiter
- Saturn
- Uranus
- Neptune
- Pluto
- Asteroids
- Comets
- Meteoroids, meteors, and meteoroids
- Space travel

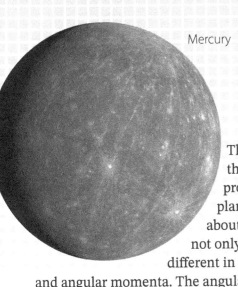

Mercury

The sun and all of the objects rotating about the sun are the **Solar System**. Diversity is probably the best way to describe it. Each planet, moon, asteroid, and every other object about the sun is uniquely different. They are not only different in appearance, they are also different in size, magnetic fields, temperatures, and angular momenta. The angular momentum is *mvr*, which is the tendency to keep rotating. This is where many of the earlier chapters you studied come together. We are very blessed today in having access to so much information about our Solar System. Astronomy had a major reformation when the *Voyager* spacecraft explored the planets and their moons. Astronomy books were rewritten. Many felt that they knew what to expect because of their expectations from evolutionary theories. When the data started to come in from these spacecraft, the mood became one of sit down and shut up. Mouths were stopped because some of the most unlikely things came to light. God gave the skills to men and women to conduct these flights and get the data back to earth to bring Him glory. It was an exciting time. It was called the Grandest Tour.

When measuring distances in the Solar System, the astronomical unit is helpful. The **astronomical unit** (AU) is the average distance between the earth and the sun — 93,000,000 miles.

Distance from Earth to the sun
93,000,000 miles (AU)

The **planets** usually come to mind first in thinking of our Solar System. They are relatively large objects that orbit the sun. They do not emit their own light but reflect light from the sun. Mercury is the planet closest to the sun. Images of Mercury were sent back to earth from the *Mariner 10* spacecraft. Radio signals from the spacecraft obeyed the inverse square law, as they spread out traveling through space. This was the craft that tested Einstein's General Theory of Relativity when the sun's gravity bent the radio signals from *Mariner 10* as it followed Mercury around the sun away from earth. Being closest to the sun, Mercury is seen when it is on our side of the sun. From earth, it always appears very close to the sun. Its temperature varies from −300°F on the side away from the sun to +800°F on the side facing the sun. Mercury has no atmosphere but does have a curious "outgassing" of helium. God has an interesting sense of humor. A day on a planet is defined as a rotation where it goes from noon to noon or midnight to midnight. A year is the time it takes a planet to revolve around the sun. A year on Mercury is 88 earth days and a day is 2 Mercurial years. That is from noon to noon. When Mercury revolves around the sun once, it has rotated only halfway around on its axis. It takes another year for it to rotate around the other half. You would have a weird calendar on Mercury. This is a great example of God's unique creative power. It does not even suggest a common form of evolutionary formation of planets. Mercury has no moons. It's average distance from the sun is 0.387 AU (Astronomical Units). Like the other planets, it has an elliptical orbit around the sun as per Kepler's first law of planetary motion.

The International Space Station is seen in silhouette as it transits the sun

Venus

The second planet from the sun is Venus, whose size is very close to that of earth. It was once thought that Venus was a sister planet to earth and that it could even possibly support life. The *Magellan* spacecraft studied the terrain of Venus with radar because of its very thick cloud cover. When Venus appears in our morning sky, it is called the morning star (even though it is not a star). When it appears in the evening sky, it is called the evening star. Later the Soviet Union (before it collapsed) sent two landers to the surface of Venus which melted shortly thereafter. It managed to send back some pictures of its yellowish rocky terrain. The atmosphere of Venus is about 100 times thicker than ours and is mostly CO_2. This means that heat and ultraviolet radiation from the sun penetrates the cloud cover and heats the surface. The hot surface radiates infrared radiation that is trapped by the atmosphere. This makes the surface of Venus about 870°F, where it can melt lead and sulfur. Volcanoes spew out molten sulfur lava. Before Venus was better understood, there were legends that a super race of women called Amazons ruled the planet and killed off all of the male population. It is about 0.723 AU from the sun. Mercury and Venus have phases like the moon because they are closer to the sun than earth. Venus has a retrograde rotation (backward in contrast to the other planets) period of 243 earth days. This is backward, and very slow.

The third planet is our earth. Mercury and Venus have no moons and we have one — sort of. It is actually called the earth-moon system because our moon is larger than a typical moon (satellite) with gravitational influence on the ocean tides. Our atmosphere is about 1% that of Venus, so we cannot expect the same greenhouse effect as Venus. Our atmosphere is mostly N_2 and that of Venus is mostly CO_2. We have an atmosphere, temperature, liquid water, and other qualities that support life, which is no accident. When God created earth, He made it to be inhabited and it became a home when He created our early parents to live here. It is amazing to think about what the new heavens and earth will be like.

Earth

Mars

Mercury, Venus, Earth, and Mars are called the **terrestrial or rocky planets**. The others — Jupiter, Saturn, Uranus, and Neptune are the **gas giants** and Pluto, in 2006, was no longer called a planet because of its small size but a **dwarf planet**.

Mars is called the red planet because of the iron content in its surface rocks and dust. It is about $\frac{1}{2}$ the size of earth and has 2 moons — Phobos and Demos. Its period of rotation is 1.03 times that of ours and the tilt of its axis is very close to ours. To have seasons, a planet needs to have an appreciable (but not

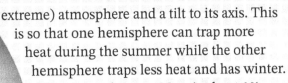

extreme) atmosphere and a tilt to its axis. This is so that one hemisphere can trap more heat during the summer while the other hemisphere traps less heat and has winter. The atmosphere on Mars is about 1% of ours. There are polar ice caps with frozen water and frozen CO_2 (dry ice). Even though Mars is about half the size of earth, it has the largest canyon in the Solar System that is as long as from the east coast to the west coast of the United States. It also has the largest (not active) volcano in the Solar System. Unique indeed!

Jupiter is by far the largest planet. Some have even said that it is almost big enough to be a dwarf star. It has more mass than all of the other bodies in the solar system combined except for the sun. It has many moons. Each of the moons has its own

Jupiter

unique qualities. This is the Creator crying out from the rooftops that He is there. The four closest to Jupiter and the largest are called the Galilean moons because they were discovered by Galileo with his early telescope. Jupiter is mostly gas. It has a great hurricane that is larger than earth that was seen by Galileo in the 1600s and is still going strong today — called the Great Red Spot. Of its moons, Io has live volcanoes spewing out molten sulfur. Considering its enormous size, it makes a complete rotation in about 8 hours. That is very fast. The *Voyager* spacecraft also found rings around Jupiter.

On September 5, 1977, *Voyager 1* was launched. On August 20, 1977, *Voyager 2* was launched. They each had separate trajectories so that *Voyager 1* arrived at Jupiter on March 5, 1979, and *Voyager 2* arrived at Jupiter on July 9, 1979. One of the astonishing finds was the live volcano on Io. This was confirmed when *Voyager 2* flew by. We discussed velocities, accelerations, momenta, and vectors in earlier chapters — these concepts in much more depth were the tools that got the *Voyagers* where they needed to be. It is like someone beginning violin lessons in contrast to someone who is a world-renown concert violinist. Remember, the most skilled individuals started somewhere. Just think, you might become a renowned physicist like Albert Einstein!

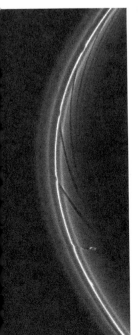

On November 12, 1980, *Voyager 1* arrived at Saturn and *Voyager 2* arrived on August 25, 1981. Saturn is the second largest planet, with an average temperature of –292°F. It has the lowest density of any planet. Isn't it rewarding that you know what *density* is from the earlier chapter? Saturn, of course, is best known by its extensive system of rings which are composed of particles of varying sizes — like gravel compared to the size of a house compared to the size of a small mountain. Saturn has a **shepherding satellite** named Prometheus. When *Voyager 1* flew by, it took pictures of what looked like part of the F ring that was braided — like you would braid a girl's hair. That was baffling. That would not happen by itself with the laws of nature. It would be like a thrown baseball swirling when it was thrown straight. When *Voyager 2* came by, they deliberately looked for the braided ring. This time the ring was braided farther down and there was a small satellite that was following the ring, going in circles around it. The gravity from the satellite was braiding the ring. That was even more astounding than the braids themselves.

Braided ring of Saturn

Saturn

Saturn also has many moons. One of the astounding ones is Titan. Titan has an atmosphere that is mostly nitrogen like earth. But before you start making vacation plans on Titan, consider that it is extremely cold. It has rivers and streams of liquid methane (natural gas). Do not worry about an explosion or fire because it is just too cold. Can you imagine the fun that science fiction writers would have with this one? I personally find science fiction boring. It cannot hold a candle to the amazing things in reality.

After Saturn, *Voyager 1* changed its trajectory by about 90° and went out of the plane of the Solar System. On January 24, 1986, *Voyager 2* arrived at Uranus, the third-largest planet, but because of its distance little was known about it up to this point. It is a large, turquois, gaseous planet. The gaseous planets have different colors from their different gas compositions. It was discovered in 1781 by William Herschel. Using the law of gravity devised by Newton, Herschel realized that there must be a massive object that slightly changed the orbit of Saturn. He looked where he expected to find this object and there was the culprit — Uranus. *Voyager 2* found fainter rings and a shepherding satellite like the one around the F ring of Saturn. Uranus also has retrograde rotation like Venus. It makes a complete rotation in about 19 earth hours. That is a great amount of angular momentum for an object this size. Some have hypothesized that Uranus evolved like the other planets with normal rotation. Then something hit it to cause it to rotate backward. Something big enough to give it that much angular momentum would have knocked it out of orbit.

Uranus

Uranus' moon Miranda has a very unusual system of valleys and cliffs that, from a distance, give the appearance of what is called a chevron. It looks like a huge number 7. In my own crazy sort of way, I thought how would God sign His name on a planet or moon. In Scripture, the number 7 stands for perfection. Just a thought. Some of its moons, like Ariel, have a surface that looks like something hit it hard and fluid from underneath oozed up and froze. When an object shows evidence of changes, it is called **tectonic**.

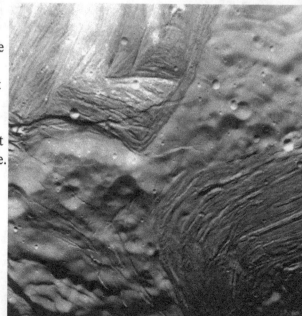

This image of Miranda, obtained by Voyager 2 on approach, shows an unusual "chevron" figure and regions of distinctly differing terrain on the Uranian moon.

Neptune

Neptune is the 8th planet from the sun. it is a blue planet with a large, blue hurricane, similar to Jupiter's Giant Red Spot. It has 4 dark rings with shepherding satellites and 2 large moons and at least 6 smaller moons. Its moon, Triton, has a retrograde rotation. It is the coldest object known in the Solar System with a temperature of −236°C or 37K. If you need to, go back and review the Kelvin scale and its relationship to absolute zero (the 3rd law of thermodynamics).

Pluto was not explored by *Voyager 2* because its orbit is not in the plane of the other planets. It was later explored by the New Horizons spacecraft. It has been found to have several moons.

It was thought at one time by some that if an object had moons it had to be a planet. That did not hold up, however, because the asteroid Ida has a moon.

Between Mars and Jupiter is a large **asteroid** belt. Asteroids are considered to be **minor planets**. They are rocky but smaller than planets. In considering the pattern of the distances between planets, the asteroid belt is where a planet would be considered to be. At one time, it was proposed that a planet was there at one time but busted apart and formed the asteroids. When combined, the asteroids would not be enough to form a planet. There are also other regions in the solar system where asteroids are found. Some that come closer to earth are monitored.

Pluto

Comets, called dirty snowballs, have large orbits that are way out in the outer fringes of the solar system. Once in a while, earth's gravity brings one in closer and it comes in, goes around the sun, and goes back to the outer portion of the solar system. Every time one goes around the sun, it loses some of its material. After several passes around the sun, it should no longer be a comet. Many comets are spotted by an amateur astronomer doing backyard observations. Some have comets named after them. If the solar system is billions of years old, we should have run out of comets. It was proposed that there is a large region just outside the solar system, called the Oort Cloud, where more comets reside, though it has not yet been proven by direct observation.

Asteroid Ida and its moon

The nucleus of a comet is ice and small particles of rocky material. When a comet comes closer to the sun, the sunlight melts some of the ice and knocks loose some of the rocky material. Gases released, along with dust, form a large region around the nucleus called the coma. The coma glows from fluorescence. Sunlight pushes some of the dust out into a dust tail. Solar wind, composed of charged particles from the sun, pushes charged particles away from the comet called an ion tail. This is why there are 2 tails on a comet. These tails are only there when the comet is closer to the sun, because they are formed by sunlight and solar wind. This is also why the tails always point away from the sun — not always behind the comet. If you have been able to see a comet in the night sky with the unaided eye, you are blessed. Some ancient peoples thought that comets were bad omens of evil spirits. They are part of the wonder of the universe that God created. The more you learn to appreciate the blessings and beauty that God created, the less tempted you will be to think that it just came about.

Meteoroids are tiny particles in space that are usually the remains from comet tails. When we have a meteor shower, earth passes through a region of space where a comet left particles of dust and small rocks behind from their dust tails. Most meteoroids are the size of a grain of sand. Once in a while a larger one enters our atmosphere and burns up or even hits the ground. **Meteors** are meteoroids that enter our atmosphere. They are also called "shooting stars." They glow because as they are coming through our atmosphere, the friction with the air produces bright glowing tunnels of hot air. Most meteors burn up in the atmosphere. Once in a great while, a meteor is large enough to survive and hit the ground. Then it is called a **meteorite**. There have been a few close calls with meteorites.

In 1984, there was a meteorite that landed in Antarctica that was thought to have originated from Mars. There were some carbonate minerals deposited in fractures of the meteorite. Some proposed that ancient bacteria-like living organisms on Mars helped form the carbonate deposits the way some have been known to do on earth. It was thought that about 15 million years ago, a large comet or asteroid crashed into

The two distinct tails of Comet C/2020 F3 (NEOWISE) captured near Villanovaforru (Italy).

Mars and knocked a chunk into space, and about a million years later crashed into earth. Notice that several assumptions were made here. For a while, it spurred interest in space exploration, but then fizzled and sort of disappeared. The microscopic carbonated deposits could be formed from the ice in the nucleus of a comet. Unmanned landings on Mars have never found life or conditions where life could survive.

There are so many more things that could be said about our solar system. I hope that this discussion has somehow aroused your curiosity to learn more. There are some good resources at the Answers in Genesis website on astronomy. As a biologist, I enjoy it as a hobby. Sometimes I just like to go out and watch the night sky. The earth that God prepared for us is the only known place in the solar system — or universe for that matter — that could support life.

I worked on a team at the University of California at Berkeley studying the feasibility of long-term space travel. One of the big obstacles is carrying enough food, water, and oxygen for the trip. We designed a recycling system where biological wastes went into a tank with bacteria that broke it down — sort of like a chemical toilet. Then algae would use the released nutrients as food and multiply. The algae took CO_2 and converted it into O_2 by photosynthesis. My astronaut was a mouse — a cute little guy. I could not train him to harvest the algae and turn it into food, so I had to prepare a diet for him and feed him from the outside. His little space capsule was sealed for over a week with the only O_2 produced from the algae. It worked great. For it to work, the algae had to be converted to food. Algae is very bitter. I took some and sent it to the home economics school at UC Davis. I figured that if they could offer a Ph.D. in home economics, they would know how to make a cake. One pesky problem is that sugar would weigh too much to take on the space craft for a long space trip, so they could not use sugar. The algae cake I got back was terrible. Before I would volunteer to be an astronaut to Mars, I would have to sample the food first.

A bright meteor streaks across the night sky near O'Leary Peak, Arizona.

LABORATORY 26

The Night Sky

REQUIRED MATERIALS

- *The Stargazer's Guide to the Night Sky*
- Flashlight
- Red cellophane (from craft store)
- Rubber band
- Small notebook to write in

Introduction

The moon is a magnificent structure gracing the night sky. Our word *month* comes from the word *moon*. A satellite is held by the gravity of the object it revolves around but is small enough that its gravity does not affect the larger body. Our moon definitely affects earth, especially in the ocean tides. As the moon goes around the earth, we say that it goes through different phases. When it is between earth and the sun, it is called a New Moon. We cannot see it because the light from the sun is reflected from the moon back to the sun. When the moon comes a quarter of the way around earth, it is called the First Quarter Moon. It is recognized by its right side reflecting light to us from the sun. When the earth is between the sun and the moon, it is called a Full Moon. It is usually enough above or below earth's shadow so that it reflects light back to earth. When the moon is three-quarters of the way around the earth, it is called the Third Quarter or Last Quarter Moon. It is recognized because its left side is reflecting light to us from the sun.

Constellations are groups of bright stars in different regions of the sky. Ancients used to think of them as representing animals or people (such as Orion the hunter). As the earth revolves about the sun, different regions of the sky are visible opposite the sun. This is why we have seasonal constellations. If you spend several hours observing the night sky, you will notice that some constellations set in the west and others come up in the east because earth is rotating from west to east.

The sun is a star that is part of a larger group of stars called the Milky Way galaxy. Planets revolve around the sun independent of the other stars. Their orbits can be shorter closer to the sun and much longer farther from the sun. The seasonal constellations are consistent from year to year, but the planets vary greatly in their positions in the night sky from year to year.

Purpose

This lab exercise is designed to provide experience observing and becoming acquainted with the night sky. After you finish this exercise, it would be good if you went out a couple of times a month and record the positions of many of the constellations, planets, and moon phases each time. This way you can see their annual patterns. This goes way beyond the scope of this course, but that is not a bad thing.

Procedure

observe

1. In your notebook, write down the date and time, where you are at (city, state, zip code) and the weather conditions.

2. Spend some time looking over *The Stargazer's Guide to the Night Sky*. There is a chapter on astronomy with the unaided eye, as well as sections on viewing the night sky with binoculars or telescopes. Notice the phases of the moon. Look over the constellations so that you can recognize some that are available when you go outdoors. You do not have to memorize the guide, but it will greatly enhance your experience to be acquainted with it.

question

3. With a parent's permission, go to the Internet and type in "What planets are visible tonight?" to get an idea as to what to look for and when. *Note*: Many smart phones have an app available to help you identify the constellations and planets.

research/hypothesis

4. Choose an evening that is not too cloudy. Go outside just after dark and spend some time looking at the sky. If you are in an area that has light pollution, you can still do this. The fainter objects will not be visible, but the brighter ones should be. Sometimes this may actually be helpful because you can be overwhelmed by the enormous number of stars when you have a clear, very dark sky. Depending upon what is visible when you go out, look for the moon, planets, and constellations.

experiment

5. Do this without binoculars or a telescope. Look to the north and see if you can see Ursa Major (the big bear or

the big dipper). The two stars on the end of the bowl of the big dipper point to the north star (Polaris) that is at the end of the handle of Ursa Minor (the little dipper or little bear). Polaris is named by its position, not its brightness. It is not very bright. Most of the stars in the little dipper are not very bright, so do not worry if you do not see them. Use the guide and see what else you can see using the big dipper as your starting point. If you have someone available who knows the sky fairly well to help, that would be good. If the moon is out, describe it and what phase it is in.

analyze

6. You can use a flashlight with red cellophane held over the lens with a rubber band. Your eyes will adjust when you have been out for a while. Your eyes dilate, allowing more light in so that you can see better in the dark. Red light will not affect your eyesight. If you use white light, your eyes will no longer be adjusted to the dark.

conclusion

7. You can write down notes in a small notebook or you can have someone write down what you tell them.

CHAPTER TWENTY-SEVEN

THE UNIVERSE

OBJECTIVES

At the conclusion of this lesson the student should have an understanding of

- Light years
- Parsecs
- Stars and galaxies
- Angular distances
- The Hertzsprung – Russell diagram
- Binary and variable stars
- Novas, supernovas, neutron stars, black holes
- The red shift and expanding universe

San Francisco, CA to Paris, France 8,948 km

"To whom then will you liken Me, or to whom shall I be equal?" says the Holy One. Lift up your eyes on high, and see who has created these things, who brings out their host by number; He calls them all by name, by the greatness of His might and the strength of His power; not one is missing (Isaiah 40: 25–26).

Thus says God the Lord, who created the heavens and stretched them out, who spread forth the earth and that which comes from it, who gives breath to the people on it, and spirit to those who walk on it (Isaiah 42:5).

When we look into the night sky, we cannot begin to see all that the Lord has made. God created the heavens and the earth. He upholds them with His wisdom and might. Through these many chapters, you have been studying the natural laws that all of creation obeys. They most certainly do not represent the absence of God as many have claimed. He created them and enforces them.

We have spent considerable time on mechanics, the study of vectors, distance, velocity, acceleration, rotational motion, and momentum in the realm of our everyday lives. Here on earth we measure distance in meters and miles. That works, even in distances around the world. But when we get to looking into the skies, we encounter much greater distances. Would it be practical to measure the distance between San Francisco and Paris in centimeters? That would be silly. Even meters may be too small. We would probably use kilometers. The moon is about 250,000 miles (400,000 kilometers) from earth. It would take light $1\frac{1}{3}$ seconds to reach the moon traveling at 186,000 miles per second (300,000 kilometers per second). The sun is about 93,000,000 miles (150,000,000 kilometers) from earth. It takes light about 8 minutes to go from the sun to earth. When we turn the discussion to stars, the distances jump astronomically (bad pun). It takes light more than four earth years to reach us from our closest star α - Centauri. Using miles or kilometers for this distance is about as silly as using centimeters for the distance between San Francisco and Paris. The unit of length used here is the **light year**, which is the distance it takes light to travel in one earth year. The phrase *earth year* is used because a year is different for different planets since the time it takes a planet to revolve once around the sun. A light year, is about 6 trillion miles or 10 trillion kilometers. The Andromeda galaxy is about 2,000,000 light years away. God stretched out the heavens indeed!

Another unit that is used for distances to stars and galaxies is the **parsec**. This is the only direct measurement of the distance to stars. There are several other good methods that are based upon patterns seen in the properties of stars and galaxies. The term *parsec* means a parallactic shift of one second of arc. Here is where you would be tempted to memorize a phrase without a clue as to what it means. But instead, let's break it down. A parallactic shift comes from the idea of a parallax. Hold the thumb of your right hand out straight in front of you. Close you left eye and look at your thumb with your right eye. Now close your right eye and look at your thumb with your left eye. Go back to your right eye and then back to your left eye. Your thumb will appear

to go back and forth even though it is not moving. It is as if you are moving back and forth over the distance between your eyes. When you look from the right, your thumb appears to be to the left. When you look from the left, your thumb appears to be to the right. If you take a picture of a closer star in January and then again in July, it will appear in a slightly different position in relation to the background stars. Your observations will be from different angles of earth's orbit around the sun. A parallax is where something appears to move because the observer moved. Trigonometry can be used to calculate the distance to the star using the triangles as shown below (*Diagram 27.1*).

There are 360 degrees around a circle and 180 degrees from one horizon to the other. From earth, the sky appears as a half circle of 180 degrees. The distance measured from the horizon to straight overhead (the zenith) is 90 degrees. One degree breaks down into 60 minutes and each minute is 60 seconds. If the star appears to move 1 second as we go halfway around the sun, it is called a parallactic shift of 1 second of arc. The phrase "of arc" means around the arc of a circle. The farther away a star is, the smaller the parallactic shift. The closer a star, the larger the parallactic shift. The parallactic shift of α - Centauri is less than 1 second of arc. That means that it is greater than 1 parsec away. Parallactic shifts have only been detectable with the development of modern telescopes and digital photos of stars and their backgrounds. A single parsec is equivalent to 3.26 light years, 206,265 AU, and about 19.2 trillion miles. You can see how light years or parsecs are preferred units for measuring distances to star.

The difference between stars and galaxies need to be clarified. A star is a very large heat- and light-emitting object. Our sun is a star — an average star at that. If our sun were a supergiant star, it would be so large that we would be inside of it. A galaxy is a large group of stars rotating about a common center of gravity. They usually consist of up to 100 billion stars. There are over 200 billion galaxies in the universe. That is a lot of stars! They have to be great distances from each other to keep from annihilating each other. God stretched out the heavens indeed!

If you are looking up at the night sky, how would you know how far apart different objects appear to be? Most of our time is spent looking with telescopes and such, so we need a hand method to indicate positions and distances. For example, if there is a bright object near the north star Polaris, how would you tell someone else where it is or how would you indicate its position for your own notes. If you hold up the tip of your little finger, it is 1° across. If you hold up at arm's length your 3 middle fingers, it is 4°. if you hold up your fist, it is 10° from the outer edge of your thumb to the outer edge of your finger. If you hold up your outstretched hand, it is 18° across your hand. You can hold up your hand to the sky and see that the bright object is the length of the distance across your fist from Polaris, you would say that it is 10° from Polaris. This is called the **angular distance**. It is not the true distance between objects in the sky, but rather the distance that they appear to be apart to you. The object you are describing appears to be on an arc, so we use the term *angular distance*.

Diagram 27.1
Exaggerated parallactic shift

Do you recall that we used the word *angular* when discussing motion in a circle in earlier chapters?

The angular size of the sun and the moon is about 0.5°. This means that when we have a total solar eclipse (the moon comes between the earth and the sun) that the moon perfectly blocks the sun. The sun is much larger than the moon, but the sun is also much farther away than the moon. But the size and distance of the sun and moon are such that when we look at them, they perfectly match each other. When this happens, the outer atmosphere of the sun, the corona, appears around the edges of the moon. Normally, the sun is so bright, we do not see the corona.

Total eclipse in the city of Gorbea, Chili in December 2020.

The study of stars relies upon the physics of optics because all we see of a star is the pinpoint of light in the sky. First of all, the pinpoint of light contains the full spectrum of electromagnetic radiation given off by the star. That extends all the way from lower energy radio waves to the high energy x-rays and gamma rays. The elements in a star can be identified by the spectra of the elements that it emits. For example, this is the emission spectrum of hydrogen.

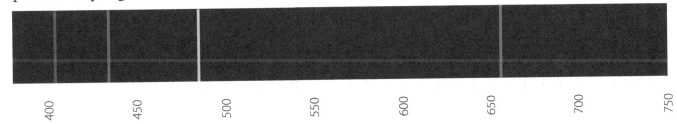

When this pattern appears, hydrogen is present. Hydrogen atoms each have one electron. The electron goes to higher energy levels and then returns to lower levels, giving off the energy as light. The different energy levels allow for different wavelengths of light to be emitted. Helium atoms each have 2 electrons with different energy levels than a hydrogen atom. The emission spectrum for hydrogen was discovered in light from our sun. This pattern had not been seen for any element known here on earth. This was the first discovery of helium. Its name is from the Greek word *Helios* for the sun. It can be said that helium was first discovered on the sun. It has since become known as the second-most abundant element in stars. The percentage of each element is determined by the intensity of the color bands in the emission spectrum. This led to the understanding that stars are made up of about 90% hydrogen and about 10% helium and trace amounts of other elements.

Hydrogen emission spectrum

The temperature of a star is determined by the most intense color given off by a star. In the color spectrum, red has the lowest energy and violet has the highest. Our sun appears yellow because out of the colors emitted, yellow is the brightest. Yellow is in the middle of the visible spectrum (red, orange, yellow, green, blue, violet), and yellow stars have a more average temperature. The hottest stars are blue and the coolest stars show as red in the sky.

Star Temperature Based on Star Color (in Kelvins)

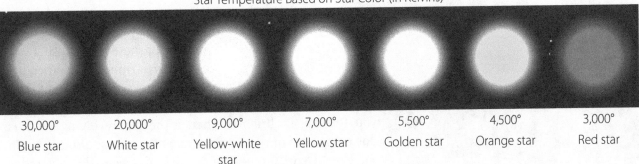

30,000°	20,000°	9,000°	7,000°	5,500°	4,500°	3,000°
Blue star	White star	Yellow-white star	Yellow star	Golden star	Orange star	Red star

Chapter 27

Now how can we determine the size of a star from the pinpoint of light? Before we can answer that question, we need to consider some vocabulary. The first is the term **apparent magnitude** of a star. This is how bright a star appears to the unaided eye. Numbers are assigned to the magnitude with lower numbers indicating brighter stars. A star with an apparent magnitude of 2 is 10 × brighter than a star of apparent magnitude 3. A star with apparent magnitude 1 is 100 × brighter than a star of apparent magnitude 3. At first, objects like planets, our moon, and the sun were not on the scale. Then, because of their greater brightness, they were assigned negative numbers that are below 0. From the apparent magnitude, the **absolute magnitude** can be determined. This is the magnitude of a star as if all stars were 10 parsecs away. This eliminates the problem of closer stars being brighter and stars farther away being dimmer. This is done mathematically.

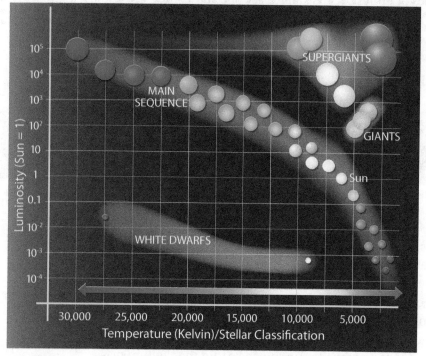

Diagram 27.1

The luminosities are plotted on a graph with the temperatures of stars. The luminosity is the amount of energy used by a star. It works like the absolute magnitude. This is called a Hertzsprung-Russell diagram named after its developers (*Diagram 27.2*).

Very dim stars with high temperatures are considered to be smaller stars (dwarfs). Stars that are very bright with low temperatures are considered to be giants. Dim stars with low temperatures and bright stars with high temperatures are considered to be average stars (called **main sequence stars**). To see the reasoning here, consider the weight and power of cars. Lightweight powerful cars are race cars. Heavy cars with low power are gas guzzlers. Heavy cars with lots of power and lightweight cars with low power are average cars.

Most stars are **binary stars**. They occur in pairs, and revolve together around a common center of gravity. Sometimes, one star is a giant and the other a dwarf. There are double binaries where two pairs revolve around a common center of gravity. Several stars in the handle of the Big Dipper are double binaries. Stars are not just thrown out there. They fit patterns of artistic design. Orderly working complexity demonstrates intent and will. Not random events.

Some stars appear to grow dim and then grow brighter, then go dim again and bright again. They keep repeating this pattern. These are **variable stars**. Their magnitudes have regular periods. The period is the time for one cycle from bright to dim and back to bright again. Some of these are eclipsing binaries. When one star in the pair is in front of the other, it eclipses (covers) the other star. This makes the pair appear dimmer. Then when they revolve so that they are both facing you, they appear brighter. Then as they are revolving the other one comes in front of the first one, they appear dimmer again. They revolve with repeating precision — again demonstrating design and having actually been placed there with precise timing. Now, why don't we see that happening when we go out and look at the sky. The stars are so far away that the binary appears as one star. These have been observed with powerful telescopes.

Some variable stars are **novas**. A nova is an explosion on a star. When a giant star is next to a dwarf, the dwarf may suck (by gravity) some of the gas off from the giant. The gas from the giant swirls and forms a thicker layer around the dwarf. The heat and weight of the added gas eventually causes the gas to explode off the dwarf, making it very bright. Then the dwarf, who has not learned his lesson, sucks more gas from the giant and does it all over again. The Hubble Space Telescope has helped a great deal in observing these phenomena. There is a lot going on out there. God is very creative. Most of the vast universe is not designed for life like that of Earth. It is designed to show us God's power and glory.

NASA image (Feb. 24, 2012) of the binary star system Eta Carinae.

A remarkable event that has been witnessed is a **super nova**. Stars, like our sun, have **hydrostatic balance** where the outward pressure from hydrogen fusion explosions in the core is balanced by the pressure of gravity pulling in toward the center, keeping stars from collapsing from the enormous gravity or exploding from the thermonuclear explosions within. Sometimes, for whatever reason, the outward pressure from within is overpowered by the gravity and a star collapses. This immense pressure causes the star to heat up to enormous temperatures and explode. The material exploding off from the star form planetary nebula — huge clouds of gas and dust that move out from the star. There is a core remnant left over in the middle. Scripture quoted at the beginning of this chapter stated that God named every star at creation and that none are lost. How does that fit with a super nova? The star is still there as a remnant. A smaller remnant will be a dwarf star. An

The Crab Nebula is what remains of a supernova.

intermediate remnant is a **neutron star**. This is where the remnant is large enough that its gravity causes it to collapse where protons combine with electrons to form neutrons. This is a smaller, though still massive star. They are also called pulsars because when they collapse, angular momentum causes the particles to swirl inward like in a tornado. This causes the neutron star to spin, sending out radio signals at 180° angles. These radio signals go out at regular periods. When they were first detected, it was thought that only intelligent life could produce such regular radio signal. They were called LGMs (little green men) because, it was thought that alien life was sending us signals. As time went by, they realized that these were neutron stars.

If the remnant is larger, it would have enough gravity to pull everything, including light, into the collapsing remnant called a **black hole**. The term *black hole* refers to the absence of light. Around the black hole is a region where the gravity is so intense that nothing can escape. This is called the event horizon. They are detected as a binary pair with an invisible partner. It is like two ghosts dancing where one is visible and the other is not. As well, as material spirals in toward the event horizon, the great velocity produces enormous amounts of friction. The heat generated releases x-rays that are thrown out at 180° angles — like the radio waves from a neutron star. They appear as invisible sources of great amounts of x-rays.

The spiral galaxy NGC 253, lies about 11.5 million light-years away. A new instrument captured an intricate whirlpool of dust, where violent star formation may be occurring around a supermassive black hole.

Some of the most distant objects in the universe are some of the brightest. These are called **quasars** — quasi stellar objects. One idea is that they could be massive black holes that are sucking in large numbers of stars, perhaps even galaxies.

Distant galaxies

An interesting phenomenon grew out of the studies of the emission spectra of stars. A **red shift** was noticed for stars and galaxies. This is where all of the bright lines in the emission spectra are moved over toward the red color. Red has less energy than the other colors, so that means that the wavelengths of the spectral lines were a little longer, indicating that they had a decrease in their energy. How could that happen? Remember that when a source of sound is coming toward you, its wavelength gets a bit shorter because the source of the sound has less distance to travel. As the source of sound goes away from you, its wavelength is stretched out (the Doppler Effect) because it has farther to travel. Sound is much slower than light. But if a source of light were moving away from you at a speed closer to the speed of light, the increased distance between you and the source of light could make a difference. If these stars and galaxies are moving away from us fast enough, the wavelengths of light could be stretched out a bit toward the decreased red side of the spectrum. In science, we make observations and then try to come up with hypotheses to try to explain them or come up with a principle. That is what happens here. The red shifts are real. So, why are they there? When the pitch of the sound of a train whistle decreases as the train is going away, it is because the speed of sound is not too different from the speed of the train. If the speed of a star moving away is not too different from the speed of light, it can produce an effect similar to the Doppler Effect. In 1929, Edwin Hubble saw a relationship between the red shifts of galaxies and their distances. The greater the red shift, the greater the distance to the galaxy. This is called the Hubble relation. How do you test such an idea since you cannot go out there where the galaxies are? You check to see if it is consistent with other observations and conclusions. This led to the idea that the universe is expanding. Here it gets a little (maybe a lot) crazy. What does it expand into? It cannot be space, because space is part of the universe. A good answer is God. You could call this one of the outer frontiers of science.

As better telescopes are employed, more astounding unbelievable things will be discovered. An all-powerful God is hard for the human mind to grasp. But looking out into the universe brings us closer and closer to the reality of God's immense power, creativity, and glory.

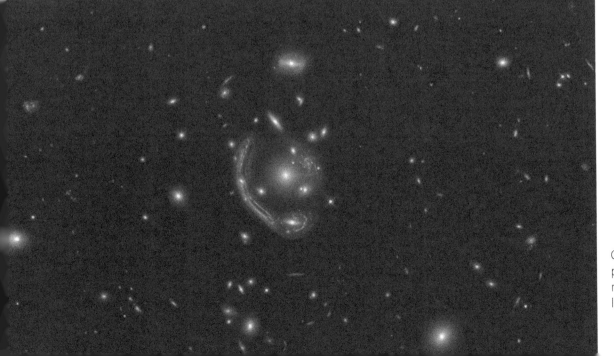

God's creative power fills the night sky with light.

LABORATORY 27

Constellations and Planets

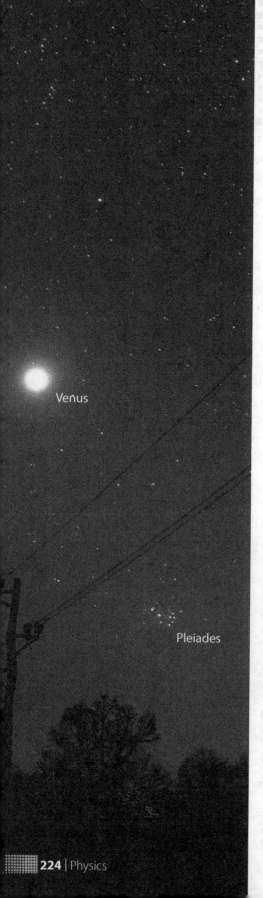

REQUIRED MATERIALS

- *The Stargazer's Guide to the Night Sky*
- Flashlight
- Red cellophane
- A rubber band
- Binoculars
- Small notebook to write in

Introduction

You may need to be a bit flexible as to when you do these observations because of weather conditions. If you live in a cold climate, cold dark nights are actually the best for observing, even though your body may not think so. When the ground is warm from the day, warm air near the ground rises as convection currents in wavy patterns distorting what you see. This is why stars twinkle at night. If the stars are not twinkling, that is good. You will see clearer images. Use the binoculars (if you have them) after you have looked with your unaided eyes. Sometimes if you look at stars out of the corner of your eye, you will see clearer images. This is because the light is being focused on the sides of the retina where the rods (black and white light receptors) are more abundant.

You are looking in the four quadrants of the sky — north, east, south, and west. The sky is usually clearest straight up because you are looking through less atmosphere. Also, the sky appears the haziest near the horizon because you are looking through more atmosphere. Sodium vapor streetlights (orangish-brown color) are the best for night observing. They do not penetrate as much and are easier on your eyes.

When you first go out to observe, you have to let your eyes adjust. The pupils of your eyes dilate (open wider) to allow more light in. Do not use the light from your cell phone, as your eyes will have to readjust afterward. After your eyes adjust, the sky will appear brighter. This is okay because more light from the sky is hitting your retina.

Purpose

This lab exercise is to provide an acquaintance with the night sky. Probably the best possible laboratory is free. Nothing can compare with the real thing. I have had many students begin lifetime hobbies that have greatly enriched their lives starting with just a brief exposure to God's magnificent glory in the heavens. Also, when tragic times come in life, going out into the night refreshes the stability of God when you see that the universe itself is still fully under God's control.

Procedure

 observe

1. This lab is to be performed out in the night sky. Try to find a place with minimal light pollution. Nevertheless, you can still do a lot with some light pollution. Use the red cellophane over the flashlight, as it is much easier on your eyes. Record what you observe.

2. In your notebook, write down the date and time, where you are at (city, state, zip code), and the weather conditions.

question

3. Look to the north. Look for the Ursa Major (Big Dipper). Find the 2 stars of the bowl opposite the handle and look in the direction from the top of the bowl. They point toward Polaris (north star). Look in *The Stargazer's Guide to the Night Sky* and see which constellations and stars you can find around the Big Dipper. Make a sketch of the positions with labels of the constellations that you can see. Indicate any planets that might be visible.

4. Look to the east and see which constellations you can identify. Sketch and identify the constellations that you find. Indicate any planets that might be visible. This will be the darkest part of the sky because the sun sets in the west and you may still have some twilight. As you look toward the horizon, it may be more difficult to identify anything because you are looking through more atmosphere.

5. Look to the south and see which constellations you can identify. Sketch and identify the constellations that you find. Indicate any planets that might be visible.

6. Look to the west and see which constellations you can identify. Sketch and identify the constellations that you find. Indicate any planets that might be visible. This may be the more difficult direction for observing because there may still be some ambient light from twilight.

research/hypothesis

7. Look in your field guide for the location of the Pleiades. This is a beautiful star cluster. Be sure to use the binoculars. If Jupiter is out, look for the 4 Galilean moons that appear as white dots in a row near Jupiter. You can actually see them with binoculars.

 experiment

8. If the moon is out, see if you can see the craters and large flat dark areas called Maria (seas). If you see any meteors, indicate what they looked like and where you saw them.

 analyze

9. Use the notes in your notebook to write a report of your findings.

 conclusion

10. Be sure to use complete sentences and include all sketches.

CHAPTER TWENTY-EIGHT
COSMOLOGY

OBJECTIVES

At the conclusion of this lesson the student should have an understanding of

- Cosmologies
- The Genesis creation account
- Big Bang cosmology
- The star light travel problem
- Extrasolar planets

In this image from the La Silla Observatory in Chile, this star cluster, known as NGC 3293, displays God's creative glory in the night sky.

Cosmology is the study of the structure of the universe. All of physical reality is in the universe, so this study results in a view of reality called a cosmology. Abraham believed in one God who created the universe. We are told that he spoke with God and God told him that He would bless him in that his offspring would be as numerous of the stars in the sky. Can you imagine the sky Abraham was able to look at without any light pollution except maybe from his campfire? Later God directed Moses to write the Book of Genesis, describing how He created the universe. These definitely shaped how both Abraham and Moses viewed life and the world around them. We all have our cosmologies, how we view the world and life around us. It is shaped by what influences us and how we respond. What voices do you listen to? As you get older, especially after high school, you will hear voices you have not heard before. It is important how you deal with them. I had a professor at USC in graduate school that was a brilliant biologist, a renowned expert in the field of biological aging. One day when he and I were sitting on the front steps of the building where our lectures were held, he said that his father was a Baptist pastor and prayed daily for his salvation. He said that when he got up in the morning, he would write for a couple of hours, then go over to the university and give a couple of lectures. Afterward, he would eat something and begin drinking. He would get so intoxicated that he could hardly stand and would go to bed. The next day, he would do the same thing. What was his view of life? What did his cosmology look like? He said that maybe someday he would accept Christ, but in the meantime, he liked his life the way it was. There were other professors who said they were atheists and laughed at my belief. At the same time, they were brilliant scientists, but with different ways of looking at life. As a blessing, there were other professors that knew Christ and were encouraging. Why the differences? It stemmed from how they viewed God, life, and the universe. The temptations you will face when you encounter these situations are huge. Do not underestimate them. What does your cosmology look like?

Hebrew scholars will tell you that the creation and flood accounts in Genesis were written in a literal historical style that is quite different from Hebrew poetry. Many Jewish scholars in Jerusalem know that the Genesis accounts are literal — they just do not believe them. Many others today treat Genesis as if it is poetic and the universe began as a Big Bang. This means that the universe evolved over billions of years rather than being created in 6 days. Many object that a 6-day creation requires miracles — actions on God's part aside from natural laws. This reflects their view of God and their cosmology.

The Greeks believed in many gods that did not have the ability to create. Their cosmology included an eternal universe. This idea is held by some today as the steady state or continuous creation theory. When Edwin Hubble popularized the idea that the universe was expanding, they said that the universe was always expanding. If this were true, the universe would be so spread out that we should not see galaxies and **star clusters**. To overcome this, they said that in the center of the universe new matter keeps popping up and expanding outward. When the Big Bang idea was introduced, a steady state astronomer, Fred Hoyle, mocked it by calling it the **Big Bang** and the name stuck.

The Big Bang states that the universe began as a small singularity that began expanding — not exploding. An explosion moves out into space, but the Big Bang has space itself expanding outward. Part of the reasoning is that if the universe is expanding today, it would have been much smaller in the past. The theory states that the universe began about 13.8 billion years ago as a very dense state that has been expanding ever since. It was a singularity like a black hole but without an event horizon — unbounded. This is because if the equations Einstein developed for General Relativity had input with an unbounded universe, it would predict a Big Bang. This is one of the first problems with a Big Bang cosmology. In 10^{-43} second energy started forming subatomic particles in this region. Then at 10^{-36} second, the singularity undergoes an inflation where it expands by a thousand billion, billion, billion times. This is needed to begin the universe, but there is known information as to why it would occur or the source of its energy. This creates a cooling effect, protons and neutrons are formed, and radiation called the cosmic microwave background radiation (**CMB**) is released uniformly into the universe. The next phase has particles coming together to form hydrogen, helium, and lithium atoms. The universe is expanding at a slower rate from there on out. Later, stars, galaxies, planets, and koala bears form. The expansion formed a homogenous universe where everything is spread out evenly. Some serious questions include how matter would have come together to form atoms, stars, galaxies, and planets.

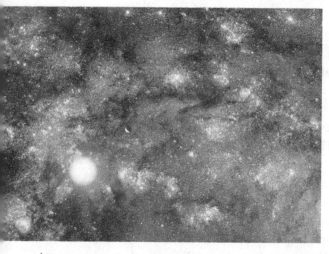

Astronomers have for the first time caught a glimpse of the earliest stages of massive galaxy construction. Dubbed "Sparky," this is a dense galactic core blazing with the light of millions of newborn stars that are forming at a ferocious rate.

This is a serious problem. Keep in mind, however, that this is the nature of science. Problems mean that more research and study is necessary. Thomas Edison made many unsuccessful light bulbs before one finally worked. Alexander Graham Bell made many unsuccessful telephones before one worked. These problems do not cause the abandonment of a theory. However, theories are discarded when an alternative is proposed that answers the concerns more fully. This has happened with so many other theories. The Big Bang cosmology has been revised many times. This is why it is important to realize that the real issue is not the problems with a cosmology, but that it is made the basis for interpreting God's testimony rather than the other way around. The biblical account is very different from the Big Bang cosmology. Many naively believe that the Big Bang is a clear concise description of the universe when it is not. The famous physicist Stephen Hawking states as well that there are several serious problems with the Big Bang. But in his mind, it is the best that there is for now. Without the Bible, that is true.

In 1965, Penzias and Wilson found microwaves coming to us from all directions from space. This was predicted as coming from the inflation stage of the Big Bang. This was called the cosmic microwave background (CMB). It is very convincing when something as precise as the CMB is predicted and later found. Because of this, many today who hold to the Day-Age theory and theistic evolution have the Big Bang as a major component of their cosmologies. Many Christians feel that Genesis 1:1 teaches the Big Bang: "In the beginning God created the heavens and the earth." But the Big Bang teaches that the earth did not come into being until about 9 billion years after the universe, yet it is mentioned in Genesis 1:1. In science, one has to remain open to the possibility that observations can have other causes. Not to do so is to stop research in its tracks. This is a problem that Niels Bohr had with his view of quantum mechanics. Many individuals who hold to a Big Bang cosmology are brilliant scientists. It is that they feel that the evidence is compelling in its favor. Therefore, Genesis must mean something else. Many unbiblical views have found their way into the church by this reasoning. You have to respect people, but at the same time, use discretion in recognizing different cosmologies. I respected my professor at USC, but did not accept his empty cosmology. Perhaps the greatest challenge to our biblical cosmology is when a Christian that we respect lets us down in a huge way. Out of bitterness and anger, we can be easily tempted to reject their testimony and accept the cosmology of a non-Christian instead.

It is important to understand other cosmologies; otherwise, you could be accepting ideas that contradict God's Word without realizing it.

A problem that is part of the young earth, literal Genesis interpretation is that of light traveling from distant stars. How did Adam and Eve see stars if stars were created on day 4 of the creation week and Adam and Eve were created on day 6. The closest star is over 4 light years away. Many stars are millions to billions of light years away. They shouldn't even be visible today. A telescope does no good if the light from the stars can't even reach you. In all honesty, that question is not answered yet, but there are a number of possible answers. Again, the description of the days of creation was given by God. The understanding of starlight is up to us and we do not do as good of a job as God. Stop and remind yourself that you do not discard God's revelation because you cannot explain it.

A more popular explanation is that of being created with age. Adam and Eve were created as adults. Plants had to be mature to be sources of food. They were also necessary to convert water and carbon dioxide into sugar and oxygen by photosynthesis. Some have proposed that God created the light already in place from the stars and galaxies. If the stars and galaxies always remained the same, the light would be a true representation of them. But that is not the case. Stars become supernovas, galaxies collide and interact with each other, black holes draw in neighboring matter, stars and galaxies. Many find this idea deceptive. If it were true, we could not trust what we see.

Another idea that has been proposed is that the velocity of light was much faster in the past. Some data from past measurements was presented but later found to be within the range of experimental error.

One that has gained a lot of popularity is the **starlight hypothesis** proposed by Russell Humphreys. He stated that if the assumption is made that the universe is bounded, then the equations for General Relativity predict a white hole in contrast to a black hole. Those who believe in the Big Bang assume the universe is unbounded. That was not exactly a black hole, because black holes are not unbounded. They have an event horizon. In a black hole, everything is drawn into the event horizon. In a white hole, everything passes out through the event horizon. In his model, as matter escaped through the event horizon, the white hole would shrink. He had earth in the middle of the white hole. Matter would go through the event horizon so fast that, according Einstein's theory of General Relativity, billions of years would go by outside of the event horizon while day 4 of creation occurred on earth inside the white hole. As more and more matter escaped the event horizon, it would shrink and disappear, giving us earth and the universe as they are today. With this cosmology, light from distant stars and galaxies would have billions of years to reach us, giving us realistic images.

Another recent proposal is the **dasha** cosmology. In Genesis 1:1 the creative act of God is described as being *ex nihilo*, meaning out of nothing. Genesis 1:11 and 12 states that the earth brought forth plants. The verb *dasha* in Hebrew is used. Some have translated this to mean that they sprouted and thrust up out of the ground. They had normal growth but rapidly. This, of course, would be a miracle — as all of creation is as well. On day 6 Adam was created from the dust of the earth in the same way. The proposal is that the stars and galaxies and their light were created in the same way. God miraculously brought the light from the

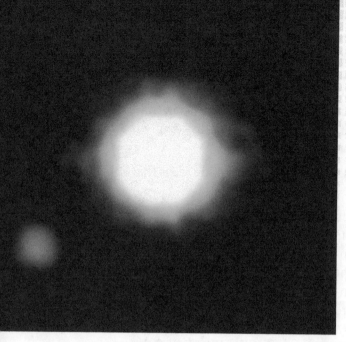

This composite image shows an exoplanet (the red spot on the lower left), orbiting a brown dwarf star. The planet is five times more massive than Jupiter.

distant stars. The difference between this view and that of created with age is that here the light actually came from the stars and galaxies.

These views would be automatically rejected by many because they involve God. Sad to say, but true, many professing Christians would rather leave God out if they at all can. Even in our personal lives, we are probably tempted to attribute God's blessings and answers to prayer to natural causes.

Extrasolar planets, also called *exoplanets*, have stirred a bit of excitement in recent years. The first to be discovered was named 2M1207 b around a star named 2M1207 in 1992. They are named by placing the lower case letter b after the name of the star. Others found around that star were named by adding the letter c and so on down the alphabet. Since then, many others have been found. That same year, 3 planets were discovered around the pulsar PSR B1257+12. This does not fit the Big Bang model of the origin of the universe because a pulsar is supposed to be formed by a supernova which would have destroyed any planet. An extrasolar planet was found around the star 51 Pegasi named 51 Pegasi b in 1995. It was found by very faint Doppler method. The planet caused slight red shifts and blue shifts as it orbited the star. A later successful method detecting **transits** of planets across stars detected many other planets. As a planet passes in front of a star, it partially blocks the light from the star. Very cleverly designed instruments have been able to detect these transits. The Kepler spacecraft, launched in 2009 and lasting until 2013, was able to study about 150,000 stars to look for planets and transits. To be successful, the planets would have to be in the orbital planes of the stars; that is, crossing in our line of sight across the star. It discovered more than 1,000 exoplanets.

The goal in evolutionary science in searching for planets is to find planets that are capable of supporting life. According to evolutionary theory, life is inevitable to evolve if the conditions are right. The first requirement is liquid water, but beyond that is a supportive atmosphere, temperature, and surface conditions. That means that the star has to be the right size and the planet has to be right distance from the star. Out of all the exoplanets discovered, none has met the proper conditions. Some have planets the right distance but the star or the planet is the wrong size. But even from an evolutionary point of view, there is no reason why life should evolve if the conditions are right. The beautiful thing from a creation point of view is that we are not obsessed with having to find life conditions but looking with amazement at the glories of the wonders of creation. The heavens bring glory to God. I believe that in these later days, God is ramping up the revelations of the wonders of His glory to win many to Him before it is too late.

Because of the topic of cosmology, it may sound like science and the Bible are at odds with each other. Cosmology falls in the category of theoretical physics, which for the most part is not tested in the laboratory like velocity, acceleration, momentum, etc. Most physics is the same for everyone. You could go through a very productive career in physics with rare instances of conflict. When going through school studying physics, it takes hard faithful work, for which you will be rewarded. In the areas that conflict with Scripture, they will want to know if you know what they are presenting. That is okay, because you should know other points of view. That is the only way that you can discuss them. At the same time, you do not have to believe everything that is presented to you.

For since the creation of the world His invisible attributes are clearly seen, being understood by the things that are made, even His eternal power and Godhead, so that they are without excuse, because, although they knew God, they did not glorify Him as God, nor were thankful, but became futile in their thoughts, and their foolish hearts were darkened (Romans 1:20–21).

His eternal power and Godhead are revealed. Indeed!

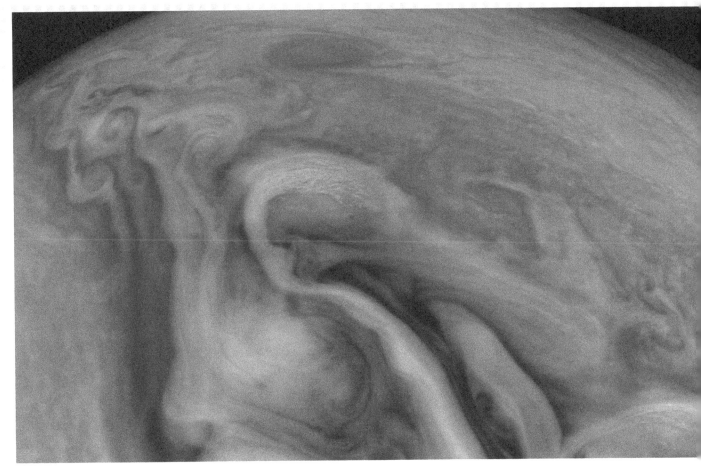

Swirling storms on Jupiter

LABORATORY 28

To Bring Glory to God

REQUIRED MATERIALS

- Flashlight
- Red cellophane
- A rubber band
- Binoculars
- Small notebook to write in

Introduction

The heavens declare the glory of God; and the firmament shows His handiwork. Day unto day utters speech, and night unto night reveals knowledge. There is no speech nor language where their voice is not heard. Their line has gone out through all the earth, and their words to the end of the world (Psalm 19: 1–4a).

Purpose

The purpose of this exercise is to see the glory of God in His creation of the heavens.

Procedure

observe

1. For this lab exercise you can do one of the following. In whichever step you choose, describe how many ways you see God glorified.

? question

2. Go outdoors and spend time observing the night sky. You can use a flashlight with red cellophane to take notes in a small notebook. Do not worry about the names of the constellations and planets unless you know them already. Spend at least an hour or more (preferably). Look into the four quadrants as you did before. Binoculars (if you have them) will enhance many objects.

research

3. Afterward, write a one-page, double spaced, 12 font paper describing your observations. Hopefully, you will develop a hobby of looking at the night sky. With time you can become very proficient at recognizing and naming many things.

hypothesis/experiment

4. This is like step 1, except go out two nights at least a week apart. Notice how much the sky has changed as earth has gone a distance around the sun in just one week.

analyze

5. Visit an observatory (if it is available) and study the exhibits, telescopes (if you have access), and the planetarium presentation.

conclusion

6. Afterward, write a one-page, double spaced, 12 font paper describing your observations.

GLOSSARY

Absolute magnitude — the brightness of a star as they would appear if they were all 10 parsecs away

Absolute zero — the absence of any thermal energy, –273 K

Acceleration — change in velocity by speeding up, slowing down or changing direction; units of meters per second squared; $\frac{\Delta v}{\Delta t}$ where Δ means "a change in"

Acceleration of gravity — the acceleration of a falling object near sea level; 9.8 $\frac{m}{s^2}$

Accommodation — the lens of the eye focusing light rays on the retina at the back of the eyeball

Accuracy — how close measurements are to true values

Alpha particles — highly energetic helium ions (He^{2+}) emitted by nuclear decay. Symbolized by the Greek letter alpha (α)

Alternating current (AC) — where an electric current alternately flows back and force – such as in household circuits

Amplitude — half the height from the crest to the trough of a wave

Angular acceleration — change in angular velocity; $\alpha = \frac{a}{r}$

Angular distance — the apparent distance across the sky measured in degrees

Angular momentum — the rotational inertia of a rotating object that must be overcome to change its rate of rotation; mvr

Angular velocity — velocity of a revolving object in terms of how many times it goes around the object per given time; revolutions per minute; $\omega = \frac{v}{r}$ in units of radians per second (2π radians in a circle)

Anode — the positive charged electrode of a power source

Antinode — the wide part of a standing wave

Apparent magnitude — the brightness of a star based upon a scale where lower numbers indicate greater brightness.

Aperture — the opening in the eyepiece of an optical instrument

Asteroid — a large rocky object revolving about the Sun that is too small to be a planet

Astronomical unit — the average distance between earth and the Sun – 93,000,000 miles

Atomic mass number — the sum of the number of protons and neutrons in an atom

Atomic number — the number of protons in an atom that identifies which element the atom belongs to

Beats — When two sound waves of slightly different frequencies are emitted together, repeated intervals of loud and faint are produced due to constructive and destructive interference. The pulses of loud sound are beats.

Beta particles — highly energetic electrons emitted by nuclear decay. Symbolized by the Greek letter beta (β)

Big Bang — a cosmology that describes the universe as beginning as a singularity and expanding outward with the formation of all the parts of the universe over a time period of about 13.8 billion years

Binary stars — pairs of stars that revolve around common centers of mass

Blackbody radiation — electromagnetic radiation emitted as colors (different frequencies) when an object is heated

Black hole — the remnant of a super massive super nova that draws everything into a singularity including light

Capacitor — two conductors separated by an insulator. Positive charge builds up on one conductor and negative charge builds up on the other conductor. When the charges exceed a particular value, the negative electrons jump across the insulator and complete the circuit. These store energy and act as timing devices.

Cathode — the negative charged electrode of a power source

Cathode ray tube — a glass tube where high voltage is passed through a low pressure gas

Centi — one hundredth in the metric system – such as a centimeter is $\frac{1}{100}$ of a meter.

Center of Gravity — the point in an object at which gravity would appear to act

Center of Mass — the average position of mass in an object; this holds true whether gravity is present or not

Centrifugal force — force opposite centripetal force according to Newton's Third Law of Motion (for every force there is an equal and opposite force); this force keeps an object revolving around another object without crashing into it and keeps earth in orbit around the Sun.

Centripetal acceleration — change in the direction of an object causing it to revolve around another object

Centripetal force — force that provides the centripetal acceleration causing an object to revolve around another

Chain reaction — a reaction that is caused by bombardment of neutrons that produces more neutrons that multiplies the reactions

Chords — intervals that overlap each other as in harmonics

Circuit — the pathway taken by electricity through a wire

Circuit breaker — opens (disrupts) a circuit when the current exceeds a specific value

Closed circuit — a pathway for the flow of electricity that has no breaks so the current can flow from the power source and back to the power source

CMB — cosmic microwave background radiation predicted from the inflationary phase of the Big Bang

Comet — a large object composed of ice, rock and dust that revolves around the Sun

Concave lens — also called a negative diverging lens. The lens bows inward.

Concave mirror — a mirror that bows inward away from the observer. It is a converging mirror.

Condensation — water comes out of air when the relative humidity exceeds 100%; areas of compression of molecules of the medium through which sound is passing

Constructive interference — a crest of a wave meets the crest of another wave or a trough meets a trough of another wave so that they add to each other

Converging lens — also called a positive lens or convex lens. The lens bows outward.

Convex lens — also called a positive converging lens. The lens bows outward.

Convex mirror — a mirror that bows outward toward the observer. It is a diverging mirror.

Cosmology — the study of the universe and a model describing a point of view of an aspect of the universe

Coulomb — the charge of 6.25×10^{18} electrons

Crest — the top of a wave circle

Critical angle — the angle of light within a transparent medium such that the light rays are reflected back into the medium. It is the principle used in fiber optics.

Current — the flow of charges (usually electrons through a wire) measured in coulombs per second which are amperes – also called amps

Dasha — the Hebrew word for "brought forth" used for the creation of life and possibly stars where light would travel to earth very rapidly through God's creative act

Daughter element — the product of a nuclear reaction

Deci — one tenth in the metric system – such as a decimeter is $\frac{1}{10}$ of a meter.

Density — mass divided by volume

Destructive interference — a crest of a wave meets the trough of another wave so that they subtract from each other

Diffraction — when a wave train meets a barrier with a narrow opening, the part the passes through the opening fans out

Diffuse reflection — when light is reflected from an irregular surface, it is scattered in many directions

Diode — a device that turns an alternating current into a direct current. Also see rectifier.

Dipole — the opposite poles of a magnet

Disorder — When disorder is greater, it is harder to find something. It means that objects can be in many different places and harder to find. When disorder increases, molecules spread out into more space.

Displacement — moving from one place to another

Direct current (DC) — electric current that flows in one direction – such as from a battery

Diverging lens — also called a negative or concave lens. The lens bows inward.

Domain — a region where many iron atoms are aligned with their magnetic north and south poles facing the same direction so that they act as a larger magnet

Doping — adding impurities to a semiconductor to make them better conductors of electric current

Doppler Effect — the frequency of sound waves increases from a source moving toward you and decreases for sound waves moving away

Dwarf planet — a large asteroid. Pluto has been reclassified into this category because it is so much smaller than the other planets.

Echo — sound waves are reflected when they strike a surface given the perception of hearing the sound twice or more

Elastic collision — a collision in which the colliding objects return to their original shape

Electric force field — a region of altered space around a positive or negative charge. It will attract an opposite charge and repel a like charge.

Electric potential — the potential of a charged object in a positive or negative electric force field. It is measured in units of energy per unit charge as $\frac{\text{Joules}}{\text{Coulomb}}$ which are volts.

Electromagnet — where an electric current (moving electric charges) produces a magnetic field that aligns the domains in a piece of iron turning it into a larger magnet

Electromagnetic induction — a moving magnetic force field produces moving electric charges (electric current) in a wire

Electromagnetic motor — a wire loop in an alternating magnetic field which causes the wire to rotate

Electromagnetic radiation — light; alternating electric and magnetic fields that induce each other moving through space

Electromagnetism — a region of interacting electric force fields and magnetic force fields

Ellipse — a modified circle that has 2 foci (plural of focus) rather than a center; looks like a flattened circle

Emission spectrum — the wavelengths of light emitted when electrons in atoms lose energy and go to lower energy levels; the emission spectrum pattern is unique to each element.

Energy — causes something to move; the ability to do work

Entropy — measure of disorder

Escape velocity — the velocity required to escape Earth's gravitational pull

Far sighted — also called presbyopia. A person can focus the image of objects at a distance but not those near.

First Law of Thermodynamics — energy can not be created or destroyed

Focal point — the point where the light rays coming through a lens come together.

Focus — an ellipse has 2 foci (the points about which an ellipse is drawn) instead of a center (like a circle)

Forced vibrations — sound waves emitted from one object cause similar vibrations in another object

Frame of reference — background against which measurements are made such as the interior of an airplane where the velocity of walking down the aisle does not consider the velocity of the airplane

Free Fall — when a falling object no longer accelerates because of air resistance

Free radical — a molecule with an atom with an unpaired electron that is highly reactive with other molecules

Frequency — the number of crests that pass a given point in a second

Fulcrum — part of a machine supporting a lever

Fundamental tone — the basic repeating sine wave of sound waves in music

Galaxy — a group of stars up to possibly 200 billions stars that move about a common center of gravity

Gamma rays — electromagnetic radiation with energy greater than X-rays, symbolized by the Greek letter gamma (γ)

Gas giant — the extremely large gaseous (in contrast to rocky) planets including Jupiter, Saturn, Uranus and Neptune

General relativity — time dilation and length contraction occurs in immense gravitational fields. It is called general relativity because immense gravitational fields exist all the time near large galaxies.

Geocentric — model of the solar system where the planets, Sun and Moon revolve around earth

Gram — metric unit of mass

Gravity — the attractive force between two objects that depends upon the mass of each object and the distance between their centers of mass

Half-life — the time it takes for a radioisotope to decay half of its radioactive material

Harmonic — a standing wave in a string or rope with 2 nodes and 1 antinode; additional meaning refers to standing waves that accompany a fundamental with multiple crests (also called overtones)

Heisenberg's Uncertainty Principle — we cannot know both the position and velocity of an electron

Heliocentric — a model of the solar system where planets revolve around the Sun;

Horsepower — unit of power in the English system;
1 hp (horsepower) = 550 foot pounds = 746 W; 1 kW = $1\frac{1}{3}$ hp

Hydrostatic balance — the outward pressure from thermonuclear explosions in the core of a star is balanced by the inward pressure of gravity

Hypothesis — a suggested principle or explanation based upon observations. Hypotheses are subject to modification or replacement to better explain experimental results.

Incident ray — a light ray hitting an object

Index of refraction — indicates how much light slows done in contrast to that in a vacuum or in air

Inertia — the property of matter to resist changes in its motion

Inertial frame of reference — the physical setting where something is measured

Impulse — when a moving object collides with a non-moving object, the momentum (mv) of the moving object becomes impulse (force × time of impact)

Index of refraction — an indication of the extent to which light is slowed down and bent as it passes through a different medium

Infrasonic — sound waves with a frequency less than 16 Hz; below that of human hearing

Integrated circuit — miniaturized enormously large electric circuit into a very small space smaller than a postage stamp

Intensity — the measured amplitude of sound waves measured in decibels (db)

Intervals — differences in frequencies between notes in a scale

Inverse Square Law — where energy, such as gravity, spreads out from its source, it diminishes as the square of its distance – such as, at a distance twice as far from its source, the force of gravity is reduced by $\frac{1}{4}$

Iridescence — as light passes through closely layered transparent layers, it is reflected by both layers producing interference patterns that allow different frequencies of light with different colors to be produced

Isotope — atoms with the same number of protons but different numbers of neutrons

Joule — metric unit of work; Newton meters

Kepler's First Law of Planetary Motion — planets revolve around the Sun in elliptical orbits

Kepler's Second Law of Planetary Motion — as a planet revolves around the Sun, it goes faster nearer the Sun and slower when further from the Sun; the radius vector of a planet orbiting around the Sun sweeps out equal areas in equal amounts of time

Kepler's Third Law of Planetary Motion — the ratio of the average radius of revolution of a planet cubed divided by its period of revolution squared is the same for all the planets (except Pluto); $\frac{R^3}{T^2}$ is the same for the planets in our solar system

Kilo — a thousand of something in the metric system – such as a kilometer is 1,000 meters.

Kilowatt hour — the use of 1,000 watts ($\frac{Joules}{second}$) of electric power in one hour, kWh

Kinetic Energy — energy of motion; $\frac{1}{2}mv^2$

Law of Conservation of Momentum — in an elastic collision, the total momenta (plural for momentum) of the colliding objects equals the sum of the momenta of the colliding objects after the collision

Laws of Nature — summary statements of many observations that always occur the same way. An example is the law of gravity that describes the force of gravity. These are not explanations, but a statement of what was observed. These are also called scientific laws.

Law of Reflection — the angle of an incident light ray to the normal equals the angle of the reflected rays to the normal

Length contraction — a unit of length (such as a meter) is shorter when going close to the speed of light or in an immense gravitational field as contrasted to that on earth

Lever — a type of machine where a plank is rotated over a fulcrum

Light year — the distance light travels in a year in a vacuum. 5,878,000,000,000 miles

Limiting Factor — a measurement that is least accurate that limits the accuracy of the overall result

Longitudinal wave — a wave that goes back and forth like a sound wave

Loudness — the perception of the amplitude of sound waves

Machine — a devise that reduces a force to do a given amount of work and/or changes the direction of the applied force

Magnetic force field — a region of altered space around a north or south pole of a magnet. It will attract an opposite pole and repel a like pole.

Main sequence stars — average size stars as determined by their placement of the Hertzsprung – Russell diagram

Mass — measure of the amount of matter in an object that determines its momentum and density

Mechanical advantage — measure of the amount that a force can be reduced by a machine; if a force of 2N is applied with a machine and 10N is needed without the machine, the mechanical advantage (MA) is 5

Mechanics — the study of motion

Melodic intervals — intervals that follow each other in a melody

Meteor — a meteoroid that is pulled into earth's atmosphere, also called shooting stars. Most of them burn up from friction in the atmosphere.

Meteorite — a meteor that is large enough to survive burning up in the atmosphere and crash into earth

Meteoroid — a piece of rocky material orbiting around the Sun. Many of them are believed to be left over from the dust tails of comets.

Milli — one thousandth in the metric system – such as a millimeter is $\frac{1}{1,000}$ of a meter.

Minor planets — the majority of the asteroids. The larger asteroids, including Pluto, are called dwarf planets.

Mirage — an object appears to be reflected from water because light rays are bent upward from the ground by warmer air and cooler air above

Model — description of something that you cannot see – such as light particles or electrons

Momentum — measure of inertia; mv

Monopole — a positive or negative charge of an ion

Moon — a satellite that revolves around a larger object

Myopia — also called near sighted. A person can focus the image of objects that are near but not those at a distance.

n-type semiconductor — a semiconductor with impurities that carry negative electrons that increases its conductivity

Natural frequency — the frequency of vibration that an object will easily vibrate at due to its dimensions, density and materials of which it is constructed

Natural radioactivity — the release of alpha particles, beta particles and gamma rays by unstable nuclei

Neap Tide — the lower ocean tides produced when the gravitational pull of the Sun and Moon are not aligned with each other

Near sighted — also called myopia. A person can focus the image of objects that are near but not those at a distance.

Neutron star — the remnant from a super nova where protons and electrons combine to form all neutrons – also called pulsars because they emit radio waves with a regular pulse

Nuclear fission — the event when the nucleus of an atom splits into two nuclei with the release of energy

Nuclear fusion — the combining of the nuclei of two atoms with the release of energy

Nucleon — a proton or neutron in the nucleus of an atom

Newton — the metric unit of force; $\text{kg}\,\dfrac{\text{m}}{\text{s}^2}$

Newton's First Law of Motion — an object at rest will stay at rest unless caused to move by a force and an object in motion will stay in motion unless there is an opposing force

Newton's Second Law of Motion — the force required to accelerate an object is the mass of the object times its desired acceleration; $F = ma$

Newton's Third Law of Motion — for every force there is an equal and opposite force

Node — the part of a standing wave that appears as a point

Normal — a line perpendicular to a reflective surface

Nova — a dwarf star draws gas from a giant in a binary pair where the added gas on the dwarf periodically explodes appearing as a variable star

Nuclear fission — the splitting of the nucleus of an atom

Nuclear force — the force that holds protons together in the nucleus of an atom

Ohm — unit of electric resistance symbolized by the Greek letter omega (Ω)

Ohm's Law — $\text{Current} = \dfrac{\text{Voltage}}{\text{Resistance}},\ I = \dfrac{V}{R}$

Open circuit — a pathway for the flow of electricity that has break so current cannot flow through it

Orbit — energy levels of electrons in Niels Bohr's model of the atom.

Orbitals — energy levels of electrons in Erwin Schrodinger's model of the atom.

Overtones — standing waves that accompany a fundamental with multiple crests (also called harmonics)

p-type semiconductor — impurities added to a semiconductor to increase its ability to receive electrons

Parallel circuit — where the resistors in an electric circuit are not connected in a row but are each connected separately to the power source

Parsec — a measure of distance to a star based upon the parallactic shift of the star, 3.26 light years

Percent Difference — measure of precision (how close measurements are to each other); % difference = $\dfrac{\text{largest value} - \text{smallest value}}{\text{average value}} \times 100$

Percent Error — measure of accuracy; how close a set of measurements are to the true value; % error = $\dfrac{\text{experimental value} - \text{true value}}{\text{true value}} \times 100$

Photoelectric Effect — light must have a certain energy level (frequency) before it can cause electrons to be ejected to start an electric current. This is sited as support for the particle model of light.

Photovoltaic effect — photons of light generate an electric current in a solar cell

Photon — a unit of light in the particle model of light

Pi — π; circumference of a circle divided by its diameter

Pitch — the frequency of sound waves in music

Planet — an object that revolves around a star that reflects light from the star

Postulate — an assumption that is made in order to propose an idea

Potential Energy — energy of an object by virtue of its position in a force field; gravitational potential energy is mgh

Power — amount of energy used over a given period; $P = \dfrac{J}{s} = W$ (watts)

Precision — how close measurements are to each other

Presbyopia — also called farsighted. A person can focus the image of objects at a distance but not those near.

Pressure — force per unit area, such as $3 \, \dfrac{N}{m^2}$

Quality — the presence of overtones (harmonics) with the fundamental of sound waves (also called timbre)

Quasar — quasi-stellar object – a very bright distant object that gives off a great deal of radio waves (uncharacteristic of stars) – perhaps massive black holes drawing in stars and galaxies

Quantum — an energy level that a subatomic particle can have. There are gaps between quanta.

RAD — radiation absorbed dose, the amount of radiation exposed to and absorbed

Radian — in angular measurements, there are 2π radians around a circle irregardless of the size of the circle

Rarefaction — areas where molecules of the medium through which sound is passing spread out and are not as dense; between areas of condensation in a sound wave

Rectifier — a device that turns an alternating current into a direct current. Also see diode.

Red shift — the emission lines from a star are all shifted toward the red end of the spectrum perhaps indicating that the star is moving away

Relativity — something is described and measured relative to its surroundings and circumstances

REM — Roentgen equivalent man,, the amount of radiation times its relative impact

Refraction — bending waves as they pass through a medium

Resistance — using electric potential energy to produce heat and/or light – such as by a bulb. Units of ohms (Ω)

Resonance — waves that undergo constructive interference with each other

Resultant — the combination of two or more vectors

Retrograde motion — planets further from the Sun from earth appear to go backward and forward again in the night sky over a period of several months

Reverberation — multiple echoes of sound waves as they bounce off from hard surfaces

Revolution — to go around an object; the earth revolves around the Sun once a year

Rogue wave — a huge wave caused by constructive interference

Rotation — to spin in a circle; earth rotates on its axis every 24 hours

Rotational inertia — a rotating object will continue to rotate at the same rate unless there is an opposing force

Rotational velocity — circular velocity of a spinning object; $\frac{m}{s}$

Scalar — a quantity that does not have direction – such as speed without indicating in which direction

Scientific method — a method of making observations; suggesting possible principles (hypotheses) or explanations for the observations; testing predictions made from the proposed hypothesis with numerous experiments; if the hypothesis is well supported by predictions that come true based upon the hypothesis, it becomes a theory.

Second Law of Thermodynamics — whenever heat is transferred from a hot object to a cold object some is lost to the surroundings

Semiconductor — have properties in between metallic conductors and insulating non-metals

Series circuit — where the resistors in an electric circuit are connected in one line

Shepherding satellites — these are small moons that revolve around some of the rings of gaseous planets. They follow the ring around the planet that by gravitational pull give the ring a braided appearance.

Simple harmonic wave (SHW) — repeating standing waves

Solar system — all of the objects that revolve around the Sun, including planets and their moons, asteroids, comets and meteoroids

Special relativity — time dilation and length contraction occur at speeds close to the speed of light. It is called special relativity because it is a rare event for something to go that fast.

Spring Tide — larger ocean tides produced by the gravitational pull of the Moon and Sun aligned together

Standing wave — where constructive interference in a string causes the wave train to appear to not be moving

Star — large light and heat emitting objects in the universe; the Sun is a star

Star clusters — groups of stars held together by gravity within larger galaxies

Star light hypothesis — proposed by Russell Humphries whereby the universe formed as a white hole. As matter left the white hole it shrank. Matter that passes through the event horizon of the white hole traveled at such high velocities that, according to Einstein's theory of General Relativity, billions of years went by while on earth, within the white hole, day 4 of creation took place. This allows for the time for light to reach earth from distant galaxies.

Superconductors — materials that have no resistance to an electric current at a very low temperature called a critical temperature

Super nova — a star that explodes producing a planetary nebula and a remnant core that becomes a dwarf, neutron star or black hole

Surroundings — everything around the system such as when heat escapes into the air, the air is the surrounding

System — this is what is being acted upon such as a pan of hot water and a thermometer

Tangential motion — the horizontal motion of an object when its centripetal acceleration (movement toward the center of a circle) becomes zero

Tectonic — a planet whose surface has been altered

Terrestrial (rocky) planet — a planet with a solid surface including Mercury, Venus, Earth and Mars

Theory — a hypothesis that makes accurate predictions in numerous experiments. A theory is subject to modification or replacement when other proposed hypotheses make better predictions

Third Law of Thermodynamics — atoms and molecules cannot reach absolute zero

Thought experiment — to envision an experiment and its outcomes that cannot immediately be done physically

Timbre — the presence of overtones (harmonics) with the fundamental of sound waves (also called quality)

Time dilation — a unit of time takes longer in contrast to an outside standard. A second lasts longer when going close to the speed of light or in an immense gravitational field than on earth.

Transverse wave — a wave train that goes up and down

Tsunami — a huge circle of water produced by an underwater earthquake

Translational velocity — $\frac{m}{s}$ in a straight line

Trough — the bottom of a wave circle

Torque (τ) — tendency of a lever to rotate about a fulcrum; τ = lever arm × perpendicular force

Transformer — a device that allows the current in a wire to induce a magnetic field that induces an electric current in another wire. If the other wire has more loops, it will have greater voltage and less current than the original wire. This is a step-up transformer. If the other wire has fewer loops, it will have less voltage and more current than the original wire. This is a step-down transformer.

Transistor — a combination of two diodes. They amplify or switch electronic signals.

Transit — a planet passes in front of a star that it orbits

Transmutation — an atom is changed from one element to another by the interaction of an alpha particle or neutron

Ultrasonic — sound waves with a frequency greater than 20,000 Hz; beyond the range of human hearing

Variable stars — stars that vary in brightness over a consistent period

Vector — a quantity that has direction – such as velocity that is distance divided by time in a given direction, 5 $\frac{miles}{hour}$ North

Volt — unit of electric potential as $\frac{Joule}{Coulomb}$

Voltage drop — the drop in electric potential that occurs as energy is used by a resistor such as a bulb

Watt — unit of power, $\frac{Joule}{second}$

Wave — energy moving from one place to another

Wave front — a line connected the crests of many light waves going in the same direction

Wave height — the distance from the crest to the height of a wave train

Wavelength — the distance between 2 crests of a wave train

Wave train — the appearance of the up and down motion over a distance

Weight — the force of gravity upon the mass of an object

X-rays — electromagnetic radiation with energy greater than ultraviolet

Photo Credits:

T-top, M-middle, B-bottom, L-left, R-right, C-center

All images shutterstock.com and getty.com unless stated.

Ark Encounter: page 10 T

Master Books: Page 179 B

NASA: Pg 5; Pg 7; Pg 14; Pg 19 All; Pg 24 T, ML; Pg 33 M; Pg 38 All; Pg 39 T; Pg 79 M; Pg 82; Pg 87; Pg 88 (Earth); Pg 136; Pg 137 @NASA-GSFC-Solar Dynamics Observatory; Pg 198; Pg 207 All; Pg 208 All; Pg 209 All; Pg 210 T, M @Voyager 2, B @JPL; Pg 211 T @Voyager 2, M, B @JPL; Pg 216 @ ESA_Hubble & NASA; Pg 221 T @ESA_Hubble & NASA, B @ESA_Hubble & NASA/N. Smith; Pg 222 T @ESA_Hubble & NASA/E. Perlman; Pg 223 T @ESA_Hubble & NASA/F Courbin, B @ESA_Hubble & NASA/S. Jha; Pg 225 @ESA_Hubble & NASA; Pg 226 @NASA, ESA, M.J. Jee and H. Ford (Johns Hopkins University); Pg 228 @NASA Goddard; Pg 230 @ESO; Pg 231

Public Domain: Pg 6 B; Pg 4 National Museum of the U.S. Navy; Pg 15 B; Pg 45 B; Pg 47 MR @US Marines/LCPL Patrick G. Pressdee, USMC; Pg 79 T @United States Space Force; Pg 99 T, M; Pg 138 M; Pg 141 B; Pg 166 @FelixMittermeier; Pg 171 B @Alex Needham; Pg 176 T; Pg 177 ML; Pg 178 R; Pg 182 @U.S. National Archives and Records Administration; Pg 183 T; Pg 196 T @United States Department of Energy; Page 213 @USDA/Coconino National Forest

Science Source: Pg 204, Pg 205

Wikipedia: Pg 23 T, B; Pg 27 R; Pg 44; Pg 85 B @Jodo; Pg 111 B @Zaereth; Pg 126 T @Reinhold MoÃàller; Pg 126-127 B @KMoore CSymphony; Pg 139; Pg 155 B @Bob Linsdell; Pg 157 B @D-Kuru Wikimedia Commons; Pg 163 B @Krystal Hamlin FLICKR; Pg 181 @Boris Lobastov; Pg 188 T @ Lankyrider; Pg 201 M; Pg 203 B @Alvesgaspar; Pg 206 @Igorzinchenko; Pg 212 @Davide Batzella; Pg 219 T @CuervoNN; Pg 222 B @International Gemini Observatory_NOIRLab_NSF_AURA; Pg 224; Pg 227 @ESO_G. Beccari; Pg 232 B @ESO_P. HoraÃÅlek;

Used under (CC BY-SA 2.0) (CC BY-SA 3.0), (CC BY-SA 4.0)

LOGIC
FROM A BIBLICAL PERSPECTIVE

This logic course will both challenge and inspire high school students to be able to defend their faith against atheists and skeptics alike.

978-1-68344-159-5

JASON LISLE

DR. JASON LISLE is a Christian astrophysicist who writes and speaks on various topics relating to science and the defense of the Christian faith. He earned his Ph.D. in astrophysics at the University of Colorado in Boulder.

TO SEE OUR FULL LINE OF FAITH-BUILDING CURRICULUM

CIVICS AND THE CONSTITUTION

AN AMERICAN VIEW OF LAW, LIBERTY, & GOVERNMENT

JAKE MACAULAY & RICKI PEPIN

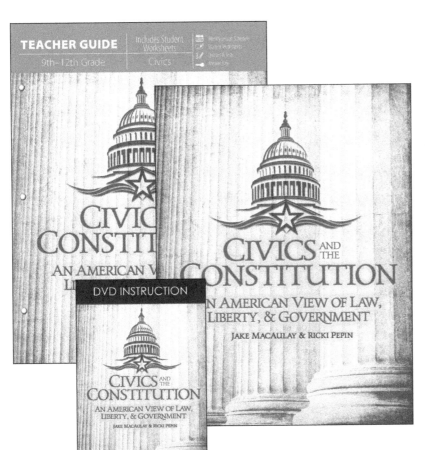

CIVICS AND THE CONSTITUTION
GRADE 10-12 *[½ YEAR / ½ CREDIT]*

Take a fast-paced and issue-focused course on the U.S. Constitution in this one semester course, complete with video lectures. Delve into the history, meaning, and daily functions of the U.S. Constitution in your life and your nation. From its original meaning to the controversies that arise from re-interpretations of it, you will understand how and why the Founding Fathers fashioned a document of God-given rights that have stood the test of time and political tinkering.

978-1-68344-186-1

VISIT **MasterBooks.com** — Where Faith Grows! — TO SEE OUR FULL LINE OF FAITH-BUILDING CURRICULUM OR CALL 800-999-3777

> **THANKS TO MASTER BOOKS, OUR YEAR IS GOING SO SMOOTHLY!**
> — SHAINA

Made for "Real World" Homeschooling

FAITH-BUILDING

TRUSTED

EFFECTIVE

ENGAGING

PRACTICAL

FLEXIBLE

We ensure that a biblical worldview is integral to all of our curriculum. We start with the Bible as our standard and build our courses from there. We strive to demonstrate biblical teachings and truth in all subjects.

We've been publishing quality Christian books for over 40 years. We publish best-selling Christian authors like Henry Morris, Ken Ham, and Ray Comfort.

We use experienced educators to create our curriculum for real-world use. We don't just teach knowledge by itself. We also teach how to apply and use that knowledge.

We make our curriculum fun and inspire a joy for learning. We go beyond rote memorization by emphasizing hands-on activities and real-world application.

We design our curriculum to be so easy that you can open the box and start homeschooling. We provide easy-to-use schedules and pre-planned lessons that make education easy for busy homeschooling families.

We create our material to be readily adaptable to any homeschool program. We know that one size does not fit all and that homeschooling requires materials that can be customized for your family's wants and needs.

VISIT **MASTERBOOKS.COM** *Where Faith Grows!* TO SEE OUR FULL LINE OF FAITH-BUILDING CURRICULUM OR CALL 800-999-3777.